T0292222

Computational Methods and Deep Learning for Ophthalmology

Computational Methods and Deep Learning for Ophthalmology

Edited by

D. Jude Hemanth

Professor, Karunya University, Tamil Nadu, India

ELSEVIER

ACADEMIC PRESS
An imprint of Elsevier

Academic Press is an imprint of Elsevier
125 London Wall, London EC2Y 5AS, United Kingdom
525 B Street, Suite 1650, San Diego, CA 92101, United States
50 Hampshire Street, 5th Floor, Cambridge, MA 02139, United States
The Boulevard, Langford Lane, Kidlington, Oxford OX5 1GB, United Kingdom

Notices
Knowledge and best practice in this field are constantly changing. As new research and experience broaden our understanding, changes in research methods, professional practices, or medical treatment may become necessary.

Practitioners and researchers must always rely on their own experience and knowledge in evaluating and using any information, methods, compounds, or experiments described herein. In using such information or methods they should be mindful of their own safety and the safety of others, including parties for whom they have a professional responsibility.

To the fullest extent of the law, neither the Publisher nor the authors, contributors, or editors, assume any liability for any injury and/or damage to persons or property as a matter of products liability, negligence or otherwise, or from any use or operation of any methods, products, instructions, or ideas contained in the material herein.

ISBN: 978-0-323-95415-0

For information on all Academic Press publications visit our website
at https://www.elsevier.com/books-and-journals

Publisher: Mara Conner
Acquisitions Editor: Chris Katsaropoulos
Editorial Project Manager: Tom Mearns
Production Project Manager: Prem Kumar Kaliamoorthi
Cover Designer: Christian Bilbow

Typeset by TNQ Technologies

Contents

viii   Contents

Contributors

R. Adarsh   National Institute of Technology, Tiruchirappalli, Tamil Nadu, India

Gadipudi Amarnageswarao   National Institute of Technology, Tiruchirappalli, Tamil Nadu, India

V.P. Ananthi   Department of Mathematics, Gobi Arts & Science College, Gobichettipalayam, Tamil Nadu, India

K. Balakrishnan   Department of Computer Science and Engineering, Indian Institute of Information Technology, Tiruchirappalli, Tamil Nadu, India

Ebenezer Daniel   Diagnostic Radiology, City of Hope National Medical Center, Duarte, CA, United States

S. Deivalakshmi   National Institute of Technology, Tiruchirappalli, Tamil Nadu, India

R. Dhanalakshmi   Department of Computer Science and Engineering, Indian Institute of Information Technology, Tiruchirappalli, Tamil Nadu, India

Poonguzhali Elangovan   Department of ECE, National Institute of Technology Puducherry, Thiruvettakudy, Karaikal, Puducherry, India

Kurubaran Ganasegeran   Clinical Research Center, Seberang Jaya Hospital, Ministry of Health Malaysia Seberang Perai, Penang, Malaysia

G. Indumathi   Mepco Schlenk Engineering College, Sivakasi, Tamil Nadu, India

Anitha J   Department of Electronics and Communication Engineering, Karunya Institute of Technology and Sciences, Coimbatore, Tamil Nadu, India

D. Jasmine David   Department of Electronics and Communication Engineering, Karunya Institute of Technology and Sciences, Coimbatore, Tamil Nadu, India

T. Jemima Jebaseeli Department of Computer Science and Engineering, Karunya Institute of Technology and Sciences, Coimbatore, Tamil Nadu, India

Mohd Kamarulariffin Kamarudin Department of Social and Preventive Medicine, Faculty of Medicine, University of Malaya Kuala Lumpur, Malaysia

S.N. Kumar Department of EEE, Amal Jyothi College of Engineering, Kottayam, Kerala, India

Malaya Kumar Nath Department of ECE, National Institute of Technology Puducherry, Thiruvettakudy, Karaikal, Puducherry, India

N. Padmasini Department of Biomedical Engineering, Rajalakshmi Engineering College, Chennai, Tamil Nadu, India

Asha Gnana Priya H Department of Electronics and Communication Engineering, Karunya Institute of Technology and Sciences, Coimbatore, Tamil Nadu, India

Ranjitha Rajan Lincoln University College, Kota Bharu, Malaysia; LUC MRC, Kuttikanam, Kerala, India

G. Santhiya Department of Mathematics, Gobi Arts & Science College, Gobichettipalayam, Tamil Nadu, India

V. Sathananthavathi Mepco Schlenk Engineering College, Sivakasi, Tamil Nadu, India

D. Selvathi Senior Professor and Head, Biomedical Engineering Programme, Department of ECE, Mepco Schlenk Engineering College, Sivakasi, Tamilnadu, India

Mohamed Yacin Sikkandar CAMS, Majmaah University, Al Majmaah, Saudi Arabia

Manavi D. Sindal Head Vitreo-Retina Services, Aravind Eye Hospital, Pondicherry, India

Bam Bahadur Sinha Department of Computer Science and Engineering, Indian Institute of Information Technology, Ranchi, Jharkhand, India

J. Sudaroli Sandana National Institute of Technology, Tiruchirappalli, Tamil Nadu, India

R. Umamaheswari Department of Electrical and Electronics Engineering, Velammal Engineering College, Chennai, Tamil Nadu, India

Ajantha Devi Vairamani   AP3 Solutions, Chennai, Tamil Nadu, India

D. Vijayalakshmi   Department of ECE, National Institute of Technology Puducherry, Thiruvettakudy, Karaikal, Puducherry, India

Alongbar Wary   School of Computer Science and Engineering, Vellore Institute of Technology—AP University, Amaravati, Andhra Pradesh, India

1

Classification of ocular diseases using transfer learning approaches and glaucoma severity grading

D. Selvathi

SENIOR PROFESSOR AND HEAD, BIOMEDICAL ENGINEERING PROGRAMME, DEPARTMENT OF ECE, MEPCO SCHLENK ENGINEERING COLLEGE, SIVAKASI, TAMILNADU, INDIA

1.1 Introduction

The primary requisite of medical diagnosis is accuracy, which has a significant effect on mitigation and treatment. Medical diagnosis is the process of identification of the presence and severity of the pathology based on the symptoms and signs. Medical diagnosis is further assisted by medical imaging techniques that are drastically improved in the past few decades. These medical imaging techniques are predominantly used for creating the visual depiction of the organs or tissues that are hidden from the human eyes by the skin. Several imaging modalities are present in the medical field. These techniques portray the structure of the retinal fundus. Vision is the prominent sense for all living beings. Therefore, eye is considered as the most precious of all sensory organs because 80% of human perception is based on what is seen. So, visual health is given much importance nowadays. Retinal examination is often done regularly. This in turn adds a responsibility of proper monitoring, analysis, and diagnosis of ocular fundus images. If any careless is done during diagnosis, it may result in delayed treatment and sometimes even loss of vision.

Glaucoma is one such dread ocular disease that mostly affects adults above 40 having a serious risk of irreversible blindness. It is the second most ocular disease to cause blindness to a huge number. WHO surveys say that it affects nearly 12 million people annually and causes 1.2 million vision losses. Glaucoma analysis demands a rich human expertise. But human analysis has an issue of intra- and interobserver variability. Thus, automatic ocular disease identification and grading are emerging nowadays. But ophthalmic imaging is a challenging job. There are several imaging technologies for capturing eye images. Among that, color fundus imaging using fundus cameras is used for imaging the rear portion of the eye. So medical diagnoses using color fundus imaging modalities are having more advantages and less risk comparing other modalities. Color fundus photography is the best method for visualizing and analyzing ocular pathologies compared to other imaging techniques.

Ocular diseases are analyzed by fundus images (rear portion of the eye) rather than lens images because the retina, optic disc, fovea, posterior pole, and macula can be visualized better in the fundus images. Medical structures like hemorrhages, optical nerve heads, blood vessels, and their abnormalities can be conveniently extracted from the fundus image. Thus, the presence and progress of most deadly ocular diseases can be analyzed from the fundus image. But human diagnosis will take considerable time and it demands expert knowledge. Therefore, automation of image analysis is needed. The automation of medical image analysis has been a topic of interest over the past 2 decades, which induced many researchers to work in this field.

1.2 Literature review

Over the past 3 decades, the automation of medical image analysis paved the attention of everyone, which led to several useful improvements in medical image processing. Several works are done by researchers in this domain. Ophthalmic imaging is a subdomain in medical imaging which is automated by the CAD process. The following are such works that automated the ocular disease diagnosis and severity grading.

The work focuses on classifying the data into two classes—diabetic macular edema and age-related macular degeneration—and also improves the network's adaptability to datasets. Duke University and Noor Eye Hospital in Tehran SD-OCT imaging are used. The reusability of the networks is improved by using transfer learning based on CliqueNet, DPN92, DenseNet121, ResNet50, and ResNext101. CliqueNet's precision, recall, and accuracy scores are comparatively higher than other networks. It represents that CliqueNet has the highest adaptability to datasets [1].

The work proposed in Ref. [2] is an automatic diabetic retinopathy classification and grading system referred as "Deep DR". The dataset used in this work is retinal fundus images from the Sichuan Academy of Medical Sciences. The first stage was a transfer learning network ResNet50 followed by the classifier Standard Deep Neural Network, which is a custom classifier. Then, the grading system works as a four class classifier (normal, NPDR, NPDR2PDR, and PDR), which utilizes ensemble learning. This strategy consistency and reproducibility for several diagnostics yielded promising results.

In [3], three Deep Neural Networks such as AlexNet, GoogLeNet, and ResNet50 performance are analyzed for image classification. Multiple datasets such as ImageNet, CIFAR10, CIFAR100, and MNIST are evaluated to prove the performance capability of the model. It is observed that increasing the training data increases the performance accuracy but also increases the complexity of the network.

In this work, the diabetic retinopathy fundus image classification using convolutional neural networks (CNNs)-based transfer learning is implemented. The publicly available retinal fundus images in DR1 and MESSIDOR datasets are used. The pretrained networks such as AlexNet, GoogLeNet, and VGGNet are evaluated for grading diabetic retinopathy (DR) in which VGGNet achieved a better performance in five classes such as DR, mild DR, moderate DR, severe DR, and proliferate DR). Two types of fine-tunings, layer-wise fine-tuning and all-layer tuning, are tested. The pretrained CNN's layer-wise will reduce the risk of overfitting and also obtain better results for small datasets [4].

This study explores the classification process using three CNNs, CifarNet, AlexNet, and GoogLeNet. The CNNs are examined for two different diseases Thoraco-abdominal lymph node and interstitial lung disease classification. This study shows that limited datasets can cause bottleneck problems. Therefore, large-scale annotated datasets can be beneficially classified using transfer learning models [5].

This work took a deep inspection on different requisites of deep learning in medical imaging. It suggested that deep convolutional neural networks (DCNNs) can auto-extract the mid and high-level features from the images. Increasing the number of iterations or epochs optimized the network parameters. When a medium-sized dataset is not available, then pretrained CNN is suggested and also the fine-tuning of the pretrained CNNs achieved better results. The challenges in medical imaging are need of a huge dataset, need of expensive medical expertise for high-quality annotation, and privacy issues in sharing the medical dataset [6].

This paper explained the processing operations to perform disease recognition using different approaches such as support vector machine (SVM), discrete cosine transform (DCT), hidden Markov model (HMM), and principal component analysis (PCA). The first step is the image acquisition followed by segmentation where the boundary of the iris is taken as circles, and they need not be cocentric. Then normalization of image is done to eliminate nonuniform illumination. Circular symmetric filter and grabber filter are used to extract features. Finally, a matching process is done by using encoding followed by hamming distance method [7].

This paper suggested a method that has a flow of region of interest (ROI) segmentation, image scaling, disc diameter calculation, cup diameter calculation, and cup-to-disc ratio (CDR) calculation from spectral domain OCT images of the Armed Forces Institute of Ophthalmology. This system uses the green channel of the pre-processed image for feature extraction. ROI extraction has two steps first extracting a circle with the center of the retina, and the next bilinear interpolation is done to improve spatial resolution and also for improving accuracy. This paper also indicates that CDR <0.5 is normal where CDR greater than or equal to 0.5 is glaucomatous and CDR increases to indicate different stages of glaucoma [8].

This work proposes a technique for glaucoma severity classification. The publicly available Online Retinal fundus Image database for Glaucoma Analysis (ORIGA) retinal fundus image dataset is used. The three steps include ROI extraction, deep feature extraction, and classification. A deformable shape model is used to localize the disc boundary, and then a rectangle with bounding box of the disc can be obtained. For deep feature extraction, AlexNet, VGG-19, and VGG-16 are used. Then, for the classification process, linear SVM is used. AlexNet gave the best performance among deep feature extractors [9].

The proposed method for an automatic glaucoma screening system is tested on retinal image (RIM) One retinal fundus image database. In preprocessing step, color channel selection and image filtering followed by illumination correction are performed. Then, there is a segmentation module that clusters the uniform super-pixels. Statistical analysis is used for feature extraction. Super-pixel classification module (cup and disc) is performed by using SVM. Then, CDR is calculated to grade glaucoma as a normal and glaucomatous image [10].

In this survey, tests are done on various publicly available retinal fundus image datasets. The segmentation process was done by two methods, optical cup and disc extraction together and optical cup and disc extraction separately. The green channel is selected for improving the segmentation speed [11]. This study shows that glaucoma assessment can be done using abnormal vision field, optic nerve damage (CDR), and intraocular pressure. Out of which CDR is the best for glaucoma assessment. This study uses the RIM-One database. CDR can be calculated using either vertical or horizontal length of both cup and disc. But vertical CDR is normally used for glaucoma assessment [12].

This work describes that glaucoma changes the optical nerve head structures such as optic disc diameter, optic cup diameter, and mean cup depth. The retinal fundus image dataset for this work is obtained from Erlangen Glaucoma Registry (EGR). The model flows as preprocessing where illumination correction is done followed by PCA for feature extraction. In the final stage, a probabilistic two-class classifier is used for glaucoma prediction [13].

This study developed an automated glaucoma risk evaluation model. EGR retinal fundus images database is used in this work. The PCA method is used to compress the extracted high-dimensional feature vectors before SVM-based classification is done [14]. In Ref. [15], Ocular Disease Intelligent Recognition (ODIR) with eight categories of diseases are classified using transfer learning approaches.

In the work [16], glaucoma grading is done using artificial intelligence in raw OCT images. In Ref. [17], deep learning-based depthwise separable convolution model is proposed to classify glaucoma from healthy images using OCT images.

In reference to these works, an automatic ocular disease classification system, which also grades glaucoma severity, is developed by implementing transfer learning using deep neural networks.

1.3 Proposed methodology

The proposed system classifies the fundus images of an eye based on eye diseases, and grading of glaucoma based on the severity of disease is shown in Fig. 1.1.

1.3.1 Dataset details

Grand Challenge Database consists of 7000 fundus images from 3500 patients (without repetition) in which 328 images correspond to age-related macular degeneration (AMD),

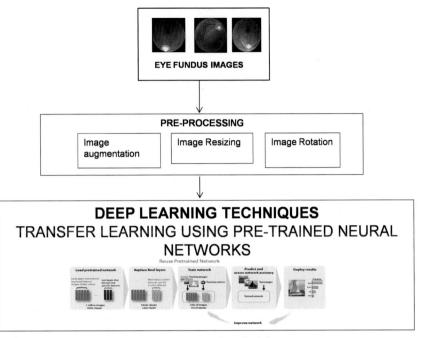

FIGURE 1.1 Proposed methodology.

424 images for cataract, 430 images for glaucoma, 206 images for hypertensive reti-nopathy, 2256 images for diabetic retinopathy, 348 images for myopia, 2280 images for normal, and others. Table 1.1 shows the distribution of the dataset.

1.3.2   Preprocessing techniques

1.3.2.1   Image augmentation
Augmentation techniques such as scaling, translation, rotation, horizontal flip, and vertical flip are applied on sample data to create multiples of new data. It is used for capturing relevant features of medical images.

1.3.2.2   Image rotation
Image rotation is the process of rotating images in either clockwise or anticlockwise direction for data redundancy purposes. It is used to improve the classification accuracy by having more number of images from different angles. For DCNN from scratch, more number of images are used to train. In this work, the image rotation is done at two different angles −45 and 45 degrees. Sample images are shown in Fig. 1.2.

1.3.2.3   Image resizing
Image resizing is the process of resizing the fundus image of an eye. A DCNN accepts input images of the same size. Because different size makes the training process a

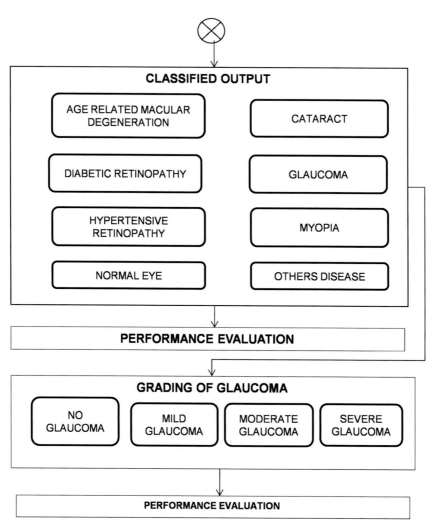

FIGURE 1.1 Cont'd

Table 1.1 Distribution of dataset.

Ocular diseases	Total images (with repetition)	Total patients (With repetition)
Cataract	424	212
Diabetic retinopathy	2256	1128
Glaucoma	430	215
Hypertensive retinopathy	206	103
Pathological myopia	348	174
Age-related macular degeneration	328	164
Others	1958	979
Normal	2280	1140
Total	8230	4115

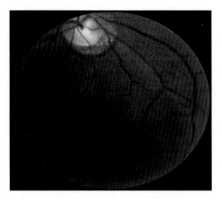
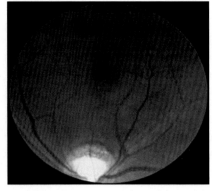

FIGURE 1.2 Sample rotated images.

tedious one and time-consuming process. Based on the pretrained neural network, the input image dimension will be varying. For that purpose, the image resizing process is used. For AlexNet, the image size must be 227 * 227.

1.3.3   Transfer learning using pretrained neural network

DCNN provides a framework for designing and implementing various pretrained models with algorithms and applications. The classification result is based on a training progress plot and confusion matrix. The classification accuracy is a more important factor in the medical field. If the classification accuracy is high, the accuracy of the classified results of eye disease is more accurate. But in this CNN method, the accuracy is a little less than the required accuracy for the medical field. Using this DCNN model, the classification accuracy is very less and also takes more days to train a model. So transfer learning using pretrained networks is chosen for better accuracy and to decrease the training time.

Transfer learning is a model trained for doing one task, which is used for doing another comparable task. AlexNet, GoogLeNet, VGGNet, and ResNet50 are some of the pretrained DCNNs used to classify ocular disease. Fig. 1.3 shows the ResNet50 architecture.

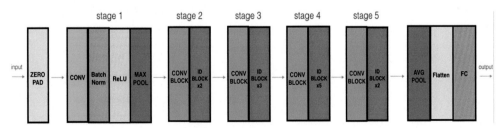

FIGURE 1.3 ResNet50 architecture.

➤ The ResNet50 model consists of five stages each with a convolution and identity block.
➤ Each convolution block has three convolution layers and each identity block also has three layers (Batch Normalization, ReLU, and MaxPool).
➤ ResNet50 has nearly 23 million trainable parameters.

Deep learning provides a promising result in medical image processing irrespective of the domain of application. Of the networks ResNet50, AlexNet, VGGNet, and GoogLeNet, ResNet50 yielded a promising result on the classification of ocular diseases with a classification accuracy of 89.9%.

1.3.4 Classified output

Fig. 1.4 shows the classified output from the deep neural network. This classification is then tested with the new test data that are not used for the training purpose. This classifier classifies some new data correctly and some new test data incorrectly because of less accuracy.

1.3.5 Grading of glaucoma

Glaucoma tremendously affects the structure of the optic nerve head also known as the optic disc. This is known as the cupping process. Therefore, optic disc is the main region to detect and analyze the presence and severity of glaucoma. As the severity of glaucoma

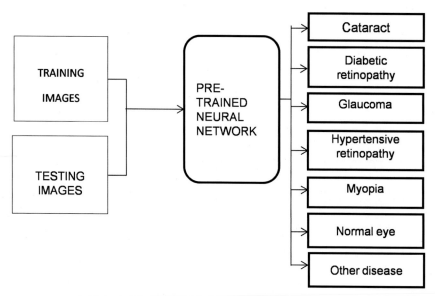

FIGURE 1.4 Classified output.

increases, the area of the nerve heads increases resulting in the cupping process. The CDR <0.5 is considered to have no glaucoma, that is, normal fundus and also 0.5 < CDR<0.7 is mild glaucoma. If 0.7 < CDR<0.8, the fundus is said to have moderate glaucoma. If 0.8 < CDR<1, it is the severe stage of glaucoma. First, the ROI, that is, cup and disc is extracted using the segmentation process. Fundus image is an RGB color image. In general, an RGB image has three channels (red, green, and blue), out of which the green channel is preferred for the segmentation process because the blue channel is considered to have low contrast. Although the vessels are clearly visible in the red channel, it consists of too much noise. Therefore, the green channel is selected for ease of segmentation. Then, the cup and disc are segmented from the background using the pixel values (green channel) in the form of a disc. Mostly cup region has a green channel intensity of 130 and the disc has 140. These regions are extracted in a disc shape. Then, the vertical cup diameter and vertical disc diameter are calculated. The ratio of vertical cup diameter to vertical disc diameter gives the CDR.

$$CDR = VCD/VDD$$

where

    CDR—cup-to-disc ratio.
    VCD—vertical cup diameter.
    VDD—vertical disc diameter.

1.3.6   Performance analysis

The neural network model is designed, and the performance is analyzed in terms of classification accuracy. Then, the grading of glaucoma is done and that is analyzed by the ophthalmologists for a second opinion.

1.4 Results and discussion

The proposed work is implemented using the computer system with the following hardware configuration and software as shown in Table 1.2.

Table 1.2 Configuration of computer and software used for experimentation.

System	64 bit OS
CPU type	Intel® core ™i6-5500 CPU
Processor	X64-based processor
Operating system	Windows 10
Clock speed	2.4 GHz
RAM capacity	8 GB
Software	Matlab 2019b

1.4.1 Classification performance analysis

1.4.1.1 Classification using CNN
Table 1.3 shows the classification accuracies achieved using CNN).

1.4.1.2 Classification using transfer learning
Tables 1.4, 1.5, 1.6 and 1.7 shows the performance results achieved for classification using GoogLeNet, AlexNet, VGG19 and ResNet50, respectively. The classification is done using an augmented dataset with 4000 images in total having 500 images in each of the eight classes. Table 1.8 shows a comparison of results obtained using transfer learning techniques.

Table 1.3 Classification result using CNN.

Number of hidden layers	Learning rate	Number of epochs	Training: testing ratio	Accuracy %
8	0.01	1	80:20	34
8	0.01	3	80:20	40.25
8	0.01	10	80:20	41.75
8	0.01	50	80:20	**50.63**

Table 1.4 Classification result using GoogLeNet.

Learning rate	Number of epochs	Training: testing ratio	Accuracy %
0.01	6	60:40	56.64
0.001	10	80:20	55.16
0.0001	15	80:20	65.31
0.0001	20	75:25	67.75
0.0001	15	75:25	**68.95**

Table 1.5 Classification result using AlexNet.

Learning rate	Number of epochs	Training: testing ratio	Validation accuracy %
0.01	1	70:30	57.17
0.01	1	80:20	53.75
0.01	10	50:50	60.88
0.001	15	70:30	65.16
0.001	15	60:40	66.57
0.001	15	80:20	68.16
0.0001	15	70:30	77.39
0.0001	15	75:25	80.24
0.0001	15	85:15	**82.36**

Table 1.6   Classification result using VGG19.

Learning rate	Number of epochs	Training: testing ratio	Accuracy %
0.01	20	75:25	66.07
0.0001	15	75:25	81.77
0.0001	15	85:15	82.49
0.0001	15	80:20	**84.25**

Table 1.7   Classification result using ResNet50.

Learning rate	Number of epochs	Training: testing ratio	Accuracy %
0.01	10	70:30	55.68
0.01	7	70:30	58.1
0.01	10	75:25	62.09
0.0001	15	75:25	87.32
0.0001	15	85:15	82.51
0.0001	15	80:20	**89.9**

Table 1.8   Comparison of transfer learning networks.

Learning rate	No of epochs	Deep CNN used	Training: testing (ratio)	Accuracy (%)
0.0001	15	Alexnet	70:30	77.39
			75:25	80.24
			85:15	82.36
		Vgg19	75:25	81.77
			80:20	84.25
			85:15	82.49
		Googlenet	75:25	68.95
			80:20	65.31
			85:15	62.59
		ResNet50	75:25	87.32
			80:20	**89.9**
			85:15	82.51

Comparing the classification results in Table 1.8, the ResNet50 with a learning rate of 0.0001, number of epochs 15, and training: testing images ratio of 80:20 achieves a greater classification accuracy of 89.9%. Fig. 1.5 shows the confusion matrix for this result.

Fig. 1.6 shows the performance comparison of the transfer learning networks.

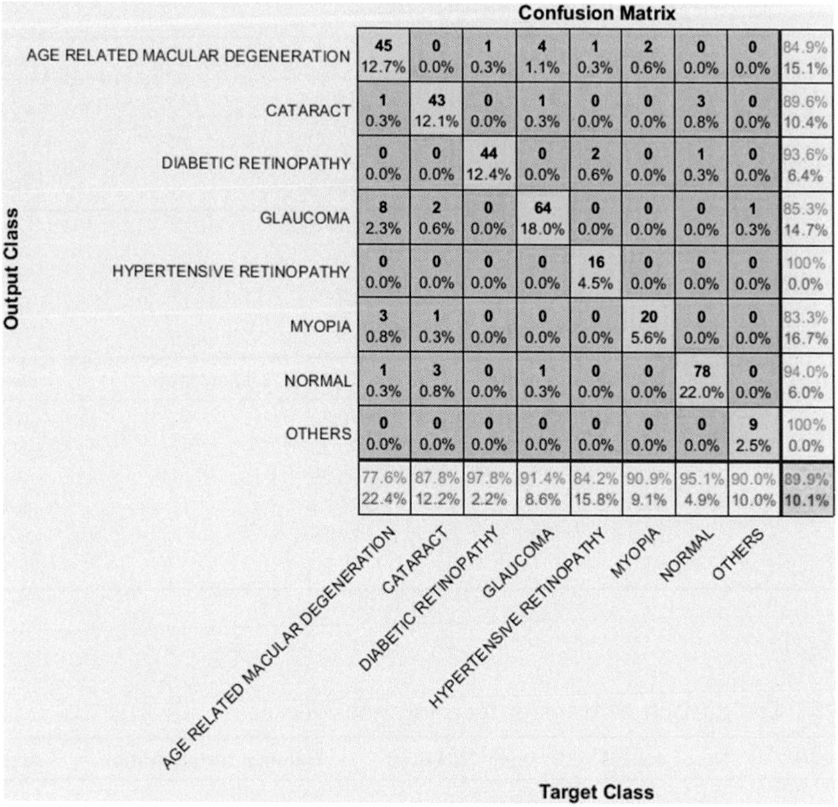

Confusion Matrix

Output Class

Output Class	AGE RELATED MACULAR DEGENERATION	CATARACT	DIABETIC RETINOPATHY	GLAUCOMA	HYPERTENSIVE RETINOPATHY	MYOPIA	NORMAL	OTHERS	
AGE RELATED MACULAR DEGENERATION	45 / 12.7%	0 / 0.0%	1 / 0.3%	4 / 1.1%	1 / 0.3%	2 / 0.6%	0 / 0.0%	0 / 0.0%	84.9% / 15.1%
CATARACT	1 / 0.3%	43 / 12.1%	0 / 0.0%	1 / 0.3%	0 / 0.0%	0 / 0.0%	3 / 0.8%	0 / 0.0%	89.6% / 10.4%
DIABETIC RETINOPATHY	0 / 0.0%	0 / 0.0%	44 / 12.4%	0 / 0.0%	2 / 0.6%	0 / 0.0%	1 / 0.3%	0 / 0.0%	93.6% / 6.4%
GLAUCOMA	8 / 2.3%	2 / 0.6%	0 / 0.0%	64 / 18.0%	0 / 0.0%	0 / 0.0%	0 / 0.0%	1 / 0.3%	85.3% / 14.7%
HYPERTENSIVE RETINOPATHY	0 / 0.0%	0 / 0.0%	0 / 0.0%	0 / 0.0%	16 / 4.5%	0 / 0.0%	0 / 0.0%	0 / 0.0%	100% / 0.0%
MYOPIA	3 / 0.8%	1 / 0.3%	0 / 0.0%	0 / 0.0%	0 / 0.0%	20 / 5.6%	0 / 0.0%	0 / 0.0%	83.3% / 16.7%
NORMAL	1 / 0.3%	3 / 0.8%	0 / 0.0%	1 / 0.3%	0 / 0.0%	0 / 0.0%	78 / 22.0%	0 / 0.0%	94.0% / 6.0%
OTHERS	0 / 0.0%	0 / 0.0%	0 / 0.0%	0 / 0.0%	0 / 0.0%	0 / 0.0%	0 / 0.0%	9 / 2.5%	100% / 0.0%
	77.6% / 22.4%	87.8% / 12.2%	97.8% / 2.2%	91.4% / 8.6%	84.2% / 15.8%	90.9% / 9.1%	95.1% / 4.9%	90.0% / 10.0%	89.9% / 10.1%

Target Class

FIGURE 1.5 Confusion matrix for the result with maximum accuracy.

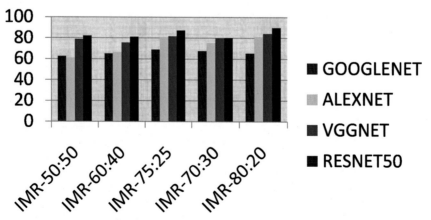

FIGURE 1.6 Performance comparison of different deep learning networks.

1.4.2   Grading results

Table 1.9 shows the grading results obtained from the expert ophthalmologists of Eye Hospital.

The methodology sample results for glaucoma grading are shown in Fig. 1.7.

The grading results from experimentation are validated by comparing with the grading results from the expert ophthalmologists of Eye Hospital, Madurai. The confusion matrix for that comparison is shown in Fig. 1.8.

Table 1.9 Glaucoma grading results by experts.

	Graded images			
Total glaucoma images	NO glaucoma	Mild glaucoma	Moderate glaucoma	Severe glaucoma
430	149	159	80	42

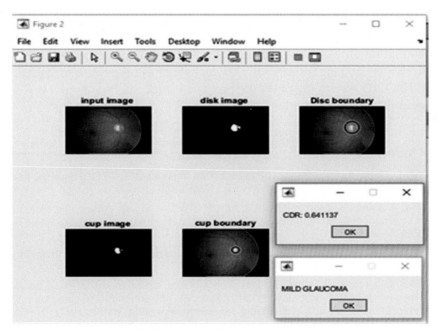

FIGURE 1.7 Sample results for glaucoma grading.

<table>
<tr><th rowspan="5">Predicted Results</th></tr>
</table>

	Mild	Moderate	Normal	Severe
Mild	123	19	15	2
Moderate	3	57	11	9
Normal	6	7	132	4
Severe	0	7	3	32

Predicted Results (vertical axis label)

Mild   Moderate   Normal   Severe

Expert Results

FIGURE 1.8 Confusion matrix for grading results.

1.5 Conclusion

In the proposed work, ocular diseases are categorized using DCNN followed by the grading of glaucoma severity. The simulation result indicates the classification result provides an accuracy of 89.9% using ResNet50. This kind of approach is helpful in diagnostics as a useful second opinion for ophthalmologists. In future work, the best network configuration will be implemented to get the highest accuracy. The methodology can also be implemented with GUI to further enhance the automation of the diagnosis. The CAD system based on the proposed work can be incorporated into a real clinical setup.

References

[1] D. Wang, L. Wang, An OCT image classification via deep learning, IEEE Photonics Journal 11 (5) (October 2019).

[2] W. Zhang, J. Zhong, S. Yang, Z. Gaom, J. Hu, Y. Chen, Y. Zhang, Automated identification and grading system of diabetic retinopathy using deep neural networks, Knowledge-Based Systems 175 (July 2019) 12−25.

[3] N. Sharma, V. Jain, A. Mishra, An analysis of convolutional neural networks for image classification, Procedia Computer Science 132 (2018) 377−384.

[4] R. Girshick, J. Donahue, T. Darrell, J. Malik, Convolutional neural networks based transfer learning for diabetic retinopathy fundus image classification, in: 10th International Congress on Image and Signal Processing, BioMedical Engineering and Informatics (CISP-BMEI), 2017.

[5] H.-C. Shin, H.R. Roth, M. Gao, Le Lu, Z. Xu, I. Nogues, J. Yao, D. Mollura, R.M. Summers, Deep convolutional neural networks for computer-aided detection: CNN architectures, dataset characteristics and transfer learning, IEEE Transactions on Medical Imaging 35 (5) (May 2016) 1285−1298.

[6] H. Greenspan, B. van Ginneken, R.M. Summers, Deep learning in medical imaging: overview and future promise of an exciting new technique, IEEE Transactions on Medical Imaging 35 (5) (May 2016) 1153−1159.

[7] L. Umesh, M. Mrunalini, S. Shinde, A review of image processing and machine learning techniques for eye disease detection and classification, International Research Journal of Engineering and Technology(IRJET). 3 (3) (Mar.2016) 547−551.

[8] T. Khalil, M.U. Akram, H. Raja, Amina jameel, imran basit, "detection of glaucoma using cup to disc ratio from spectral domain optical coherence tomography images, IEEE Access 6 (January 9, 2018) 4560−4576.

[9] Y. Wang, A. Li, Combining multiple deep features for glaucoma classification, in: IEEE International Conference on Acoustic, Speech and Signal Processing(ICASSP), Apr. 2018.

[10] N.A. Mohamed, M.A. Zulkifley, W.M.D.W. Zaki, A. Hussain, An automated glaucoma screening system using cup-to-disc ratio via simple linear iterative clustering superpixel approach, Biomedical Signal Processing and Control 53 (August 2019).

[11] A. Ahmed, R. Burman, K. Raahemifar, V. Lakshminarayanan, Optic disc and optic cup segmentation methodologies for glaucoma image detection: a survey, Journal of Opthalmology 2015 (Nov. 2015).

[12] S.E. Zohora, A.M. Khan, S. Chakraborty, N. Dey, Glaucomatous image classification: a review, in: International Conference on Electrical, Electronics and Optimization Techniques(ICEEOT), March 2016.

[13] R. Bock, J. Meier, L.G. Nyul, J. Hornegger, G. Michelson, Automated Glaucoma Detection from color fundus images, Medical Image Analysis 14 (Jan . 2010) 471−481.

[14] G.N. Laszlo, Retinal image analysis for automated glaucoma risk evaluation, Proceedings of SPIE 7497 (December 2009).

[15] N. Gour, P. Khanna, Multi-class multi-label ophthalmological disease detection using transfer learning based convolutional neural network, Biomedical Signal Processing and Control 66 (2021) 102329.

[16] G. García, et al., Circumpapillary OCT-focused hybrid learning for glaucoma grading using tailored prototypical neural networks, Artificial Intelligence in Medicine 118 (2021) 102132.

[17] A.P. Sunija, V.P. Gopi, P. Palanisamy, Redundancy reduced depthwise separable convolution for glaucoma classification using OCT images, Biomedical Signal Processing and Control 71 (2022) 103192.

2

Early diagnosis of diabetic retinopathy using deep learning techniques

Bam Bahadur Sinha[1], R. Dhanalakshmi[2], K. Balakrishnan[2]

[1]DEPARTMENT OF COMPUTER SCIENCE AND ENGINEERING, INDIAN INSTITUTE OF INFORMATION TECHNOLOGY, RANCHI, JHARKHAND, INDIA; [2]DEPARTMENT OF COMPUTER SCIENCE AND ENGINEERING, INDIAN INSTITUTE OF INFORMATION TECHNOLOGY, TIRUCHIRAPPALLI, TAMIL NADU, INDIA

2.1 Introduction

As of today, diabetes is a global illness that eventually results in total blindness. According to the 2017 International Agency for the Bar of Blindness study, there were around 422 million people with polygenic illness [1]. Diabetic retinopathy (DR) may affect one in three people with polygenic illness, and one in 10 people can lose their eyesight. Blood vessels in the retinal layer of the eye are damaged as a result of DR. Microaneurysms are formed due to the focal dilatation of weak walls that result in their formation. Exudates, which are yellow-white specks that occur when capillaries bleed, are common. One drawback of DR is that vision loss is not often apparent until the disease has advanced enough. Traditional DR screening is only useful for those with a high risk of development since there are no significant symptoms. We look for lesions and exudates in body structure images to identify DR [2].

It takes a lot of time and effort to identify DR using the conventional method, which relies on physicians to identify relevant possibilities in the body structure images. Fig. 2.1 illustrates the set of fundus images with varying degrees of diabetic retinopathy severity (from 0 to 4): (0: No DR, 1: mild, 2: moderate, 3: severe, and 4: proliferative DR). Diagnosis of diabetic retinopathy at an early stage might help those with polygenic illness to recognize the symptoms. Reduced clinical strain on tissue layer experts is expected as a result. As a result, it is easier to track the progression of the lesions. India, China, the United States, and other Asian countries account for over half of the world's diabetes prevalence. Because the numbers are projected to rise, an automated clinical detection system would be of significant use.

Computational Methods and Deep Learning for Ophthalmology. https://doi.org/10.1016/B978-0-323-95415-0.00006-1

17

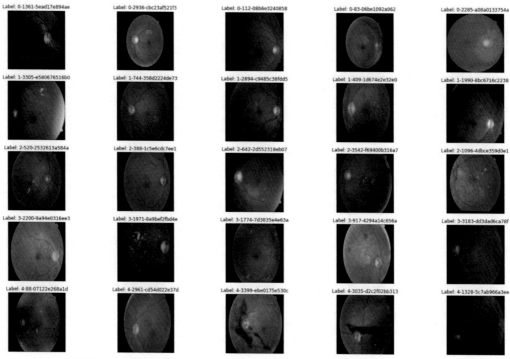

FIGURE 2.1 Severity of diabetic retinopathy: APTOS 2019 blindness detection dataset.

Microaneurysm, hemorrhage, hard exudate, and soft exudate are all indications of nonproliferative diabetic retinopathy [3]. Microaneurysms are among the most prevalent symptoms of DR, and they appear as red patches that are smaller than 125 μm in diameter and have sharp margins. It is common for hemorrhages to be bigger than 125 μm in diameter. There are two types of hemorrhages that result from capillary leaks: blot and superficial hemorrhages [4]. Plasma leakage causes hard exudate, which appears as bright yellow patches in the macular region with sharp margins. White patches on the retina are caused by swelling of nerve fibers and are called soft exudates.

Deep learning has evolved into a more durable and strong alternative that functions as a subdivision of machine learning approach. A hierarchical-based structure with multiple layers makes up a deep learning model [5]. In medical image analysis, deep learning (DL) is used to categorize, localize, segment, and identify medical images. With numerous methodologies, DL delivers more outstanding and promising outcomes in the diagnosis and categorization of diabetic retinopathy disease [6]. The model's performance improves as the amount of training data grows since both low- and high-level characteristics are seamlessly extracted and trained. Many researchers across the globe prefer convolutional neural network (CNN) over other DL approaches in medical imaging. Convolutional, pooling, and fully connected layers are the three fundamental layers of the CNN architecture. Image classification performance may be improved by using a hierarchical method to learn more complicated feature sets as well as distortion and translation characteristics in higher layers of the CNN-based image classification system. In this chapter, we investigate the application of a CNN-based technique for the detection of diabetic retinopathy. Additional tests are done using Kaggle's APTOS 2019 blindness detection dataset using a unique multilayer CNN architecture [7] and Resnet-34 [8] architecture.

The remaining sections of the chapter are structured as follows: Section 2 discusses the related background followed by different methodologies in Section 3. Section 4 presents the flow of the proposed model. Section 5 highlights the results obtained using the proposed model. The closing Section 6 concludes the chapter with a future direction.

2.2 Related background

A few years ago, the application of machine learning models in DR research was limited to identifying distinct kinds of lesions. New reasoning models have been introduced over the past few years to better understand DR gravity points. In recent years, a number of DL-based automated DR detection systems have been designed. In this section, a few of the recent research is discussed.

The presence of hemorrhages and exudates in the human body structure picture is used to identify the diabetic retinopathy level. The support vector machines (SVM) classifier is trained using images of the human body that depict the different levels of DR. When extracting the alternatives, they will descend below the classification half to determine if the supplied input image is conventional or has DR. The input check picture provided to the classifier accurately detects the amount of DR using SVM classifier training [9]. DR and diabetic macular edema may be detected in retinal fundus images using deep learning algorithms in Ref. [10]. To build the approach, the Stats Models 0.6.1 and SciPy 0.15.1 Python libraries were used. According to the scientists, a convolutional neural network may be used to create a set of local characteristics that are then aggregated into global ones. To identify local characteristics, the suggested approach has been trained. The drawback is figuring out whether or not the algorithm can be used in a clinical context. An abnormality's severity is determined using image classifiers such as multiclass SVM and K-nearest neighbor (KNN) [11]. Researchers have tried using MESSIDOR and Diabeticret DB in their research. The optic disc is eliminated once the photos have been preprocessed. The gray level co-ocurrence matrix (GLCM) algorithm is used to extract features that will be used in the classification process. It is proposed in Ref. [12] by the authors that categorization is the best way to go. A matched filter response is applied to the picture after it has been captured by the optical camera in the suggested approach. Image processing and fuzzy c-means are utilized to produce a precise image of blood vessels, making the vessels clearly apparent in the image. Features including radius, area, center angle, arc length, and 0.5 space area unit are calculated. SVM is used to classify features, and probabilistic neural networks are used to predict no diabetic retinopathy (NDR) and DR.

Data-driven categorization of DR was suggested in Ref. [13]. Colored mages of the fundus were analyzed by the algorithm, which determined if they had DR or NDR. Among the 17 features examined by the authors of the suggested approach [14] are the image's textural contrast, energy, entropy, homogeneity, area, primary axis, length of the axis, and convexo concavity. Analysis of variance (analysis of variation between groups) was employed by the authors of the proposed approach to use 11/17 features. Zago et al. [15] devised a technique for diagnosing diabetic retinopathy or non-DR fundus pictures using two CNNs (pretrained VGG16 and CNN). For training, the DIARETDBI dataset was utilized. For testing purposes, the Messidor, IDRiD, Messidor-2, DIARETDB0, DDR, and Kaggle datasets were employed. With an AUC of 0.912 and sensitivity of 0.94, the Messidor dataset produced the best results. Jiang et al. [16] proposed a model that classified the fundus image dataset using three CNNs (ReNet152, Inception-v3, and Inception-ResNet-v2). The photos were scaled, improved, and augmented before CNN training, and afterward the Adaboost approach was used to integrate them. The Adam optimizer has been used to update the network weights, and indeed the system obtained an AUC of 0.946 and an accuracy of 88.21%. Wang et al. [7] utilized the Kaggle fundus dataset and three different CNNs (Inception-v3, AlexNet, and VGG16) to identify the five-stage DR and compare the performance of the distinct CNNs. For all three pretrained models, the fundus pictures were enlarged to different sizes, yielding an accuracy of 63.23%, 37.43%, and 50.03% in Inception-v3, AlexNet, and VGG16, respectively [17]. devised a system that employed a CNN with 10 convolutional layers, eight max-pooling layers, three fully connected layers, and then a softmax classifier to categorize fundus pictures into five categories depending on DR severity levels. Color fundus photos are shrunk and adjusted. L2 regularisation and dropout approaches were employed to minimize overfitting. The model obtained 95%, 75%, and 30% of specificity, accuracy, and sensitivity, respectively. Hua et al. [18] used the DRIVE dataset pictures to locate the retinal blood vessels. ResNet-101 network which was pretrained was used to choose four feature maps, which were then combined into a single map. Before CNN processing, the fundus pictures were augmented. An accuracy of 0.951 was achieved by combining the best feature maps.

Although CNN architecture is better at spotting DR, the computational complexity is large because of the presence of a huge number of parameters. There are less parameter calculations in ResNet; therefore, this research attempts to examine how ResNet and CNN performs, yet they still have a deep network. The systematic literature review of different works done in this domain reveals the following facts:

For identifying pertaining DR in retinal fundus pictures, DL approaches yield good prediction performance

Utilizing a deep learning-based technique to identify DR in retinal fundus, pictures may help to decrease misdiagnosis.

The application of DL-based technologies lowers screening costs, increases healthcare coverage, and improves treatment outcomes.

Some of the top findings from the literature survey are demonstrated via Table 2.1.

Table 2.1 Selected features with best-obtained accuracy.

Reference	Discussion and analysis
[19]	The purpose of this study is to use a modified GoogLeNet model to categorize diabetic retinopathy and to estimate its prognosis. To ensure the model's overall accuracy is stable, cross-validation (K-fold) is done. To compare the improved GoogLeNet's effectiveness with that of other classifiers, the dataset was further trained using ResNet11 and AlexNet. The findings demonstrated the efficacy of deep learning in diagnosing and prognosing DR.
[20]	The purpose of this study is to develop an ensemble for assessing diabetic retinopathy. The proposed ensemble was composed of several components. Each component consisted of a bipartite neural network. The first section was a feature extractor, while the second section was a predictive classifier. Two models were developed in this study: binary DR classifier for performing binary classification and another for categorizing the five phases of DR (multiclassification). They employed Xception, Inception, and ResNet as feature extractors and standard deep neural network (SDNN) as a classifier for binary classification. They employed DenseNet169, DenseNet201, and Resnet50 as feature extractors and SDNN as a classifier for multiclassification. The experiment findings demonstrate the critical role and efficacy of ensembles in DR categorization.
[21]	They used an ensemble method that comprised two CNNs: VGGNet and ResNet. After training and validating these two CNN models, the results of each model were combined to create the classification model used in the ensemble training phase. According to these findings, the ensemble model was clinically acceptable in the detection of diabetes-related eye disease, including diabetic macular edema.
[22]	They designed a model to classify a series of fundus photos using a machine learning algorithm. Tuning of hyperparameters and transfer learning were used to train GoogleNet, AlexNet, ResNet, and VggNet, and then the four architectures were compared. When it comes to DR image categorization, CNNs performed very well, according to the data.

2.3 Experimental methodology

This section discusses the different deep learning architecture and data augmentation approaches used by the proposed model for classifying diabetic retinopathy.

2.3.1   Adaptive histogram equalization

In the histogram equalization output, the human eye in the original image is not able to convey appropriate information. Some information is lost owing to excessive brightness, despite the fact that the image's contrast had been increased. The reason for this is because the histogram is not restricted to the immediate area. This problem may be solved using adaptive histogram equalization. The picture is broken into tiny blocks, and the histogram is equalized for each of them in this manner. Fig. 2.2 illustrates the contrast limited adaptive histogram equalization (CLAHE) histogram equalization used to improve the contrast of the original image.

FIGURE 2.2 Original image versus CLAHE

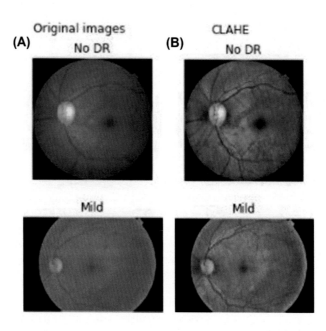

Images may be improved by using a computer image processing technology called adaptive histogram equalization (AHE). Different from the traditional technique, the adaptive method approximates multiple different images, each with its own unique set of histogram data. It then analyzes these values to reallocate the image's brightness values. Because of this, it is ideal for boosting the contrast and clarity of edges in each individual area of an image. In relatively homogenous areas of a picture, AHE tends to overamplify noise. CLAHE, a variation of AHE, avoids this from happening by restricting amplification. It is common for AHE to overamplify contrast in sections of the picture that are almost constant, as the histogram in these areas is very concentrated. Thus, noise may be enhanced in areas that are near-constant. An adaptive histogram equalization version known as "Contrast Limited AHE," or "CLAHE," [23] reduces noise amplification by restricting contrast amplification. Contrast enhancement in CLAHE is determined by the transformation function's slope in the vicinity of a pixel value. That pixel value's histogram value is directly proportional to the slope of the local cumulative distribution function (CDF). Before calculating the CDF, CLAHE clips the histogram at a predetermined value. This restricts the slope of the CDF and hence the transformation function's slope. To determine the so-called clip limit, the histogram's normalization and hence its size must be taken into consideration. Between 3 and 4, common values restrict the amplification. Instead of discarding the clipped portion of the histogram, it is better to spread it evenly throughout all histogram bins. Redistribution will result in an effective clip limit that is bigger than the specified limit and the precise amount of which varies from image to image. This may be avoided by repeating the redistribution operation iteratively till the excess is eliminated.

2.3.2   Convolutional neural network

When it comes to deep learning architectures, CNN is the most well known. In this design, each neuron is patterned in such a manner that they react to overlapping areas in the field of vision. CNNs are a significant class of learnable depiction techniques that were prompted by biological neural networks (NNs). Over the course of many years, several alternatives have been offered. The fundamental components, on the other hand, are fairly similar. CNNs are made up of a series of convolution and pooling actions that are alternated. Convolutional layers are often alternated with pooling layers to save computation time while simultaneously increasing spatial and configuration stability; the last few levels (adjacent to the output) will be fully linked to 1D layers. A feed-forward neural network may be thought of as a function "Z" of mapping data "X," which can be expressed via Eq. (2.1).

$$Z(X) = Z_L(...Z_2(Z_1(X_1, W_1)W_2)..., W_L) \tag{2.1}$$

where $L \rightarrow$ depth of NN, $X \rightarrow$ input data, and $W \rightarrow$ weight. It is important to comprehend the idea of CNNs before diving into the details of the concept. CNN is classified as a supervised algorithm, which means it is monitored. The algorithm trains using training data, which may be anything from a collection of photos in the input to a set of images with their corresponding labels in the output. Essentially, it comprises providing the CNN with pictures of the training set (X) and their corresponding targets (y), to train the network's function, $y = f(X)$ [24]. The network's function is tested with unseen pictures in an attempt to forecast their labels once the parameters of the network's function (specifically, weight and bias) have been learned. The architecture of ConvNet or CNN is illustrated via Fig. 2.3.

The method returns a five-value array representing the score (or severity of diabetic retinopathy) associated with the image's expected labels. As a result, the highest score is

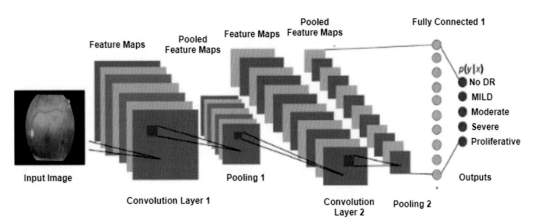

FIGURE 2.3 CNN architecture.

the projected label (or class) that should be retained for the tested image. Different roles played by different layers of CNN are discussed as follows:

- *Convolutional layer:* This is the algorithm's primary layer. It entails extracting the image's primary properties (smoothness, edge, etc.). This is accomplished by performing a series of two-dimensional convolutions of the input picture with one or more filters. Convolution is conducted concurrently on each channel of the input picture, for example, a color image contains C = 3 channels, that is RGB where "R" refers to Red, "G" denotes Green, and "B" represents Blue color. For practical purposes, the filters are adjusted to have an odd size C × F × F, for example, 3 × 5 × 5. This procedure returns a single scalar value. To further describe the input picture, the Convolution layer is repeated many times. An artificial neuron is depicted as a spatial location or pixel in the features map. The feature maps are the output of the convolution layer.
- *Maxpooling layer:* There is a spatial reduction in feature maps, and only the most important information is retained. This procedure is shown in the following image. In most cases, a 2 × 2 × 2 pooling with a stride of two provides satisfactory results for most applications. Other forms of pooling exist, of course. For example, sum pooling, median pooling, average pooling, etc.

2.3.3   ResNet

ResNet facilitates the training of more complex networks as the layers train residual functions relating to their inputs rather than training unreferenced functions. This makes the networks resistant to the vanishing gradient issue and mitigates the accuracy deterioration seen in traditional deep networks.

A direct connection facilitates identity mapping, which ensures that each successive layer has the information required to learn more characteristics. ResNets are composed of modularized designs in which building components with the same connecting form are stacked one on top of the other. These blocks are referred to as "residual units". Fig. 2.4 illustrates a basic residual block [25].

As seen in Fig. 2.4, the most critical idea is the skip link/connection. This connection is essentially an identity mapping in which the preceding layer's input is appended straight to the output of the subsequent layer. The residual function can be defined via Eq. (2.2).

$$Y(\text{Output}) = F(i) + i \tag{2.2}$$

where "*i*" denotes the input and $F(i)$ is the output from the layer. Highly complicated networks with about 100 layers may be trained using ResNets easily. The benefit of including this sort of skip connection is that any layer that degrades architectural performance will be bypassed by regularisation. As a consequence, extremely deep neural networks may be trained without the issues caused by vanishing gradients.

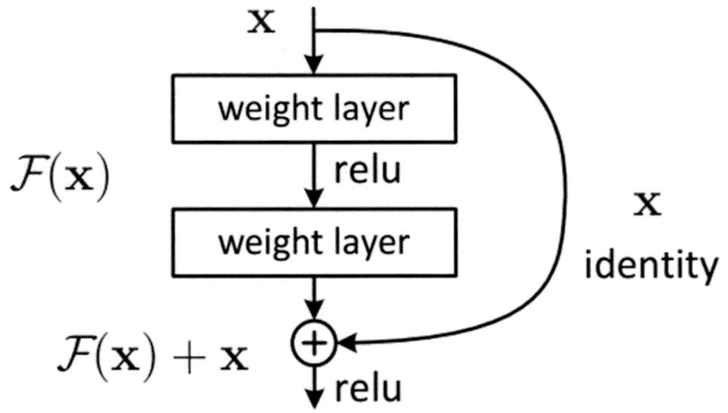

FIGURE 2.4 Basic residual block.

2.4 Proposed flow

The proposed framework can be divided into four stages as illustrated via Fig. 2.5. The first stage of the framework deals with collecting and loading the APTOS 2019 blindness detection dataset from Kaggle followed by an exploratory data analysis of the dataset. The second stage adjusts image contrast using Gaussian Blur and performs data augmentation using CLAHE. After dividing the dataset into train set, validation set and test set, the model is trained using different deep learning models in the third stage of the proposed framework. In this chapter, ResNet34 and DenseNet121 are used to train the model. Stochastic gradient descent is used as an optimizer by both models to control the generalization gap between the training and validation error. The performance of the deep learning-based models is tested in the last stage of the proposed framework by computing the training accuracy, validation accuracy, and test accuracy. Results obtained by each stage are discussed in the upcoming section of the chapter.

2.5 Results and discussion

The dataset used for the experimentation in this chapter is APTOS 2019 blindness detection dataset. Table 2.2 gives a brief description about the dataset. The dataset is quite imbalanced. The different severity categories in the training set are not in equal ratio. This encouraged us to apply data augmentation on the dataset before actually performing the classification task. The imbalanced class information present in training data can be illustrated via Fig. 2.6. The images in the training and test data comprise varied height and width. Maximum width for training set is 4288; minimum width for training set is 474; maximum height for training set is 2848; and minimum height for training set is 358.

The maximum width for test set is 2896; minimum width for test set is 640; maximum height for test set is 1958; and minimum height for test set is 480. Fig. 2.7 illustrates the

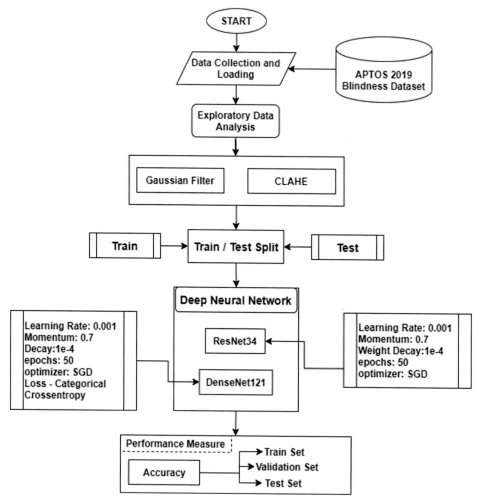

FIGURE 2.5 Proposed flow for classification of diabetic retinopathy.

Table 2.2   APTOS 2019 blindness detection dataset description.

Training set images	Test set images	Severity of DR
3662	1928	0: No DR; 1: Mild; 2: Moderate; 3: Severe; 4: Proliferative DR

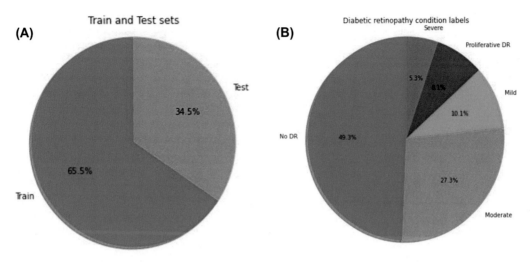

FIGURE 2.6 Severity of DR in training data.

training and test image dimensions. Gaussian Blur processes the image by passing the image through a low-pass filter. Gaussian Blur makes use of a Gaussian filter to compute the transformation of each pixel in the image. It helps in reducing the noise in the image [26]. In general, it is preferable to divide the Gaussian blur procedure into two passes: (1) The image is blurred using a 1-D kernel in vertical and horizontal direction, (2) The remaining portion of the image is blurred using the same 1-D kernel. The final effect is identical to that of single-pass convolving with a 2-D kernel but needs less computations. Fig. 2.8 illustrates the effect of Gaussian blur on different DR severity. By integrating the Gaussian function throughout each pixel's area, the precision is maintained (at a very little processing expense).

It is CLAHE's job to keep things from becoming too bright. CLAHE uses tiles, rather than the complete image, to do its analysis. CLAHE limits the contrast and hence avoids the image from becoming overbright and prevents it from losing meaningful details of the image. Fig. 2.9 demonstrates the original image versus the image after implementing CLAHE algorithm.

After preprocessing the image, the DR dataset is divided into train set, validation set, and test set. The training set is used to train the deep learning model. ResNet34 and a DenseNet model are used to perform the classification of diabetic retinopathy. Learning rate is set to 0.001, momentum is set to 0.7, weight decay is set to 1e−4, and the optimizer being used is stochastic gradient descent. The model is trained for 50 epochs. Table 2.3 demonstrates the training and validation error obtained for different number of epochs.

The total number of parameters, trainable parameters, and nontrainable parameters of DenseNet model are (7,301,189), (7,217,541), and (83,648), respectively. It can be observed from the table that maximum training accuracy obtained using ResNet is

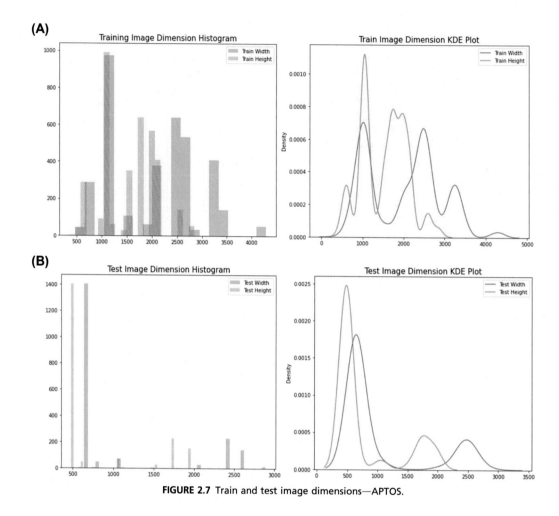

FIGURE 2.7 Train and test image dimensions—APTOS.

98.364010% and the maximum validation accuracy obtained using ResNet is 84.634825%. The maximum training accuracy obtained using DenseNet model is 98.49% and maximum validation accuracy obtained using DenseNet is 87.37%. The test accuracy obtained using ResNet and DenseNet after preprocessing the test images is 88.41% and 89.57%, respectively.

Table 2.4 demonstrates a comparative analysis of obtained results with other state-of-art models. We have compared our proposed model with modified Xception model [27], deep neural network (using pretrained VGG16 model) [29], MobileNetV2-SVM model [28], and a fused model combining CNN512 and YOLOv3 [30]. A new CNN model for DR severity categorization was built using an altered version of the Xception framework and the aggregation of deep CNN layers [27]. To accurately and efficiently classify DR severity, the suggested technique can effectively combine feature maps of various

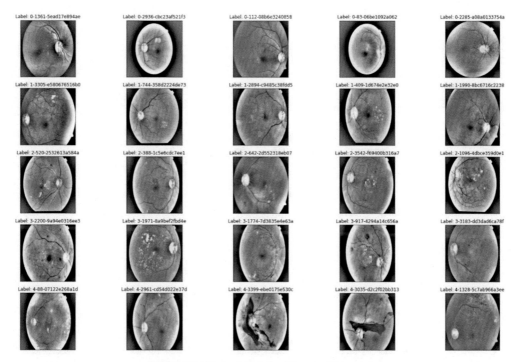

FIGURE 2.8 Effect of Gaussian filter on DR images.

depths. Min-pooling preprocessing is utilized to enhance the contrast of input pictures. It is intended that the model be trained without the use of any additional data sources at all. Additional techniques such as L1 and L2 regularization were used to deal with the dataset's highly unbalanced classes, allowing the model to be more generalizable to new, unobserved data while also reducing the likelihood of overfitting. After testing the model on the APTOS dataset, it was observed that the updated Xception deep extractor produced a much higher performance as compared to the conventional Xception framework. To conduct the DR classification in the APTOS 2019 dataset, Taufiqurrahman et al. [28] employed the MobileNetV2 architecture, which was deemed as a small-scale framework for image classification. Initialization was done using ImageNet's MobileNetV2 pretrained weights, and data augmentation and resampling were employed while training the model. By integrating the proposed model with an SVM classifier, they were able to create MobileNetV2-SVM, a hybrid and efficient deep learning model. Bodapati et al. [29] suggested a method for deep feature engineering that relies on the aggregate of features from several VGG16 convolution blocks. The experimental results reveal that when compared to hand-crafted features, a more accurate depiction of retinal pictures is provided by deep features. For DR pictures, features from pretrained models' convolutional layers provide a better representation, particularly when several convolution layers' features are combined. The experiments demonstrate

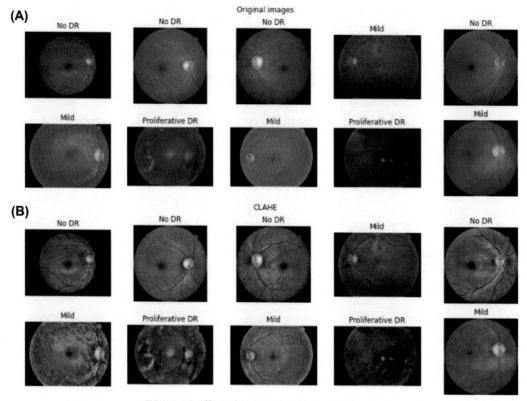

FIGURE 2.9 Effect of CLAHE algorithm on DR image.

Table 2.3   Deep learning model performance over 50 epochs.

	ResNet34	ResNet34	DenseNet121	DenseNet121
Epoch	Train Acc.	Valid Acc.	Train Acc.	Valid Acc.
1	49.394306%	39.722222%	62.21%	52.70%
5	69.957601%	68.611111%	82.48%	71.07%
10	81.223501%	78.888889%	85.94%	73.03%
15	90.854028%	78.888889%	87.79%	73.85%
20	97.092671%	77.777778%	87.99%	79.59%
25	98.001211%	78.611111%	93.00%	81.39%
30	98.394912%	79.166667%	96.15%	83.47%
35	98.326422%	81.314329%	97.49%	84.87%
40	98.323445%	83.182238%	98.77%	86.22%
45	98.364524%	84.912175%	99.21%	87.11%
50	98.364010%	84.634825%	98.49%	87.37%

Table 2.4 Performance comparison with other state-of-art models.

Reference	Approach	Accuracy
[27]	Modified Xception model	83.09%
[28]	MobileNetV2-SVM	85%
[29]	DNN	84.31%
[30]	CNN512 and YOLOv3	89.0
	ResNet34 + CLAHE	88.41%
	DenseNet121 + CLAHE	89.57%

that the suggested weighted aggregating and pooling procedures are quite beneficial for aggregating information from several blocks. Alyoubi et al. [30] proposed a model that combined CNN512 and YOLOv3. The proposed model was able to classify the DR images accurately and thus promoted an automated screening system that achieved high precision.

2.6 Conclusion and future direction

The objective of this chapter is to present a deep learning model for classifying the damages induced by DR on retinal images. To enhance the network's representational capacity, Gaussian Blur and CLAHE has been performed in the preprocessing step. Furthermore, evaluation findings on APTOS 2019 Blindness Detection dataset demonstrate that our model has good modeling and analysis characteristics, accurately recognizing different symptoms associated with DR severity levels. We plan to increase prediction effectiveness in the future by employing segmented retinal images.

References

[1] V. Jackins, S. Vimal, M. Kaliappan, M.Y. Lee, Ai-based smart prediction of clinical disease using random forest classifier and naive bayes, The Journal of Supercomputing 77 (5) (2021) 5198–5219.

[2] O. Faust, R. Acharya U, E.Y.-K. Ng, K.-H. Ng, J.S. Suri, et al., Algorithms for the automated detection of diabetic retinopathy using digital fundus images: a review, Journal of Medical Systems 36 (1) (2012) 145–157.

[3] S. Guo, Fundus image segmentation via hierarchical feature learning, Computers in Biology and Medicine 138 (2021) 104928.

[4] P. Amrelia, Pre-proliferative and proliferative retinopathy, in: Diabetic Retinopathy: Screening to Treatment 2E, ODL, 2020, p. 53.

[5] B.B. Sinha, R. Dhanalakshmi, Building a fuzzy logic-based artificial neural network to uplift recommendation accuracy, The Computer Journal 63 (11) (2020) 1624–1632.

[6] N. Asiri, M. Hussain, F. Al Adel, N. Alzaidi, Deep learning based computeraided diagnosis systems for diabetic retinopathy: a survey, Artificial Intelligence in Medicine 99 (2019) 101701.

[7] X. Wang, Y. Lu, Y. Wang, W.-B. Chen, Diabetic retinopathy stage classification using convolutional neural networks, in: 2018 IEEE International Conference on Information Reuse and Integration (IRI), IEEE, 2018, pp. 465–471.

[8] K. Oh, H.M. Kang, D. Leem, H. Lee, K.Y. Seo, S. Yoon, Early detection of diabetic retinopathy based on deep learning and ultra-wide-field fundus images, Scientific Reports 11 (1) (2021) 1–9.

[9] V. Gulshan, L. Peng, M. Coram, M.C. Stumpe, D. Wu, A. Narayanaswamy, S. Venugopalan, K. Widner, T. Madams, J. Cuadros, et al., Development and validation of a deep learning algorithm for detection of diabetic retinopathy in retinal fundus photographs, JAMA 316 (22) (2016) 2402–2410.

[10] B.S. Mankar, N. Rout, Automatic detection of diabetic retinopathy using morphological operation and machine learning, ABHIYANTRIKI International Journal of Engineering and Technology 3 (5) (2016) 12–19.

[11] J. Lachure, A. Deorankar, S. Lachure, S. Gupta, R. Jadhav, Diabetic retinopathy using morphological operations and machine learning, in: 2015 IEEE International Advance Computing Conference (IACC), IEEE, 2015, pp. 617–622.

[12] S. Sayed, V. Inamdar, S. Kapre, Detection of Diabetic Retinopathy Using Image Processing and Machine Learning, IJIRSET, 2017.

[13] R. Gargeya, T. Leng, Automated identification of diabetic retinopathy using deep learning, Ophthalmology 124 (7) (2017) 962–969.

[14] M. Rajkumar, P. Charulatha, P.H. Bindu, A. Kiruthika, Diagnosis of diabetic retinopathy using machine learning algorithms, International Research Journal of Engineering and Technology 6 (3) (2019) 7027–7030.

[15] G.T. Zago, R.V. Andreão, B. Dorizzi, E.O.T. Salles, Diabetic retinopathy detection using red lesion localization and convolutional neural networks, Computers in Biology and Medicine 116 (2020) 103537.

[16] H. Jiang, K. Yang, M. Gao, D. Zhang, H. Ma, W. Qian, An interpretable ensemble deep learning model for diabetic retinopathy disease classification, in: 2019 41st Annual International Conference of the IEEE Engineering in Medicine and Biology Society (EMBC), IEEE, 2019, pp. 2045–2048.

[17] H. Pratt, F. Coenen, D.M. Broadbent, S.P. Harding, Y. Zheng, Convolutional neural networks for diabetic retinopathy, Procedia Computer Science 90 (2016) 200–205.

[18] C.-H. Hua, T. Huynh-The, S. Lee, Retinal vessel segmentation using roundwise features aggregation on bracket-shaped convolutional neural networks, in: 2019 41st Annual International Conference of the IEEE Engineering in 400Medicine and Biology Society (EMBC), IEEE, 2019, pp. 36–39.

[19] H. Takahashi, H. Tampo, Y. Arai, Y. Inoue, H. Kawashima, Applying artificial intelligence to disease staging: deep learning for improved staging of diabetic retinopathy, PLoS One 12 (6) (2017) e0179790.

[20] W. Zhang, J. Zhong, S. Yang, Z. Gao, J. Hu, Y. Chen, Z. Yi, Automated identification and grading system of diabetic retinopathy using deep neural networks, Knowledge-Based Systems 175 (2019) 12–25.

[21] V. Bellemo, Z.W. Lim, G. Lim, Q.D. Nguyen, Y. Xie, M.Y. Yip, H. Hamzah, J. Ho, X.Q. Lee, W. Hsu, et al., Artificial intelligence using deep learning to screen for referable and vision-threatening diabetic retinopathy in africa: a clinical validation study, The Lancet Digital Health 1 (1) (2019) e35–e44.

[22] S. Wan, Y. Liang, Y. Zhang, Deep convolutional neural networks for diabetic retinopathy detection by image classification, Computers and Electrical Engineering 72 (2018) 274–282.

[23] A.W. Setiawan, T.R. Mengko, O.S. Santoso, A.B. Suksmono, Color retinal image enhancement using clahe, in: International Conference on ICT for Smart Society, IEEE, 2013, pp. 1–3.

[24] J. Qin, W. Pan, X. Xiang, Y. Tan, G. Hou, A biological image classification method based on improved cnn, Ecological Informatics 58 (2020) 101093.

[25] P. Korfiatis, T.L. Kline, D.H. Lachance, I.F. Parney, J.C. Buckner, B.J. Erickson, Residual deep convolutional neural network predicts mgmt methylation status, Journal of Digital Imaging 30 (5) (2017) 622−628.

[26] A. Haghofer, S. Dorl, A. Oszwald, J. Breuss, J. Jacak, S.M. Winkler, Evolutionary optimization of image processing for cell detection in microscopy images, Soft Computing 24 (23) (2020) 17847−17862.

[27] S.H. Kassani, P.H. Kassani, R. Khazaeinezhad, M.J. Wesolowski, K.A. Schneider, R. Deters, Diabetic retinopathy classification using a modified xception architecture, in: 2019 IEEE International Symposium on Signal Processing and Information Technology (ISSPIT), IEEE, 2019, pp. 1−6.

[28] S. Taufiqurrahman, A. Handayani, B.R. Hermanto, T.L.E.R. Mengko, Diabetic retinopathy classification using a hybrid and efficient mobilenetv2-svm model, in: 2020 IEEE REGION 10 CONFERENCE (TENCON), IEEE, 2020, pp. 235−240.

[29] J.D. Bodapati, N.S. Shaik, V. Naralasetti, Deep convolution feature aggregation: an application to diabetic retinopathy severity level prediction, Signal, Image and Video Processing 15 (5) (2021) 923−930.

[30] W.L. Alyoubi, M.F. Abulkhair, W.M. Shalash, Diabetic retinopathy fundus image classification and lesions localization system using deep learning, Sensors 21 (11) (2021) 3704.

Comparison of deep CNNs in the identification of DME structural changes in retinal OCT scans

N. Padmasini[1], R. Umamaheswari[2], Mohamed Yacin Sikkandar[3], Manavi D. Sindal[4]

[1]DEPARTMENT OF BIOMEDICAL ENGINEERING, RAJALAKSHMI ENGINEERING COLLEGE, CHENNAI, TAMIL NADU, INDIA; [2]DEPARTMENT OF ELECTRICAL AND ELECTRONICS ENGINEERING, VELAMMAL ENGINEERING COLLEGE, CHENNAI, TAMIL NADU, INDIA; [3]CAMS, MAJMAAH UNIVERSITY, AL MAJMAAH, SAUDI ARABIA; [4]HEAD VITREO-RETINA SERVICES, ARAVIND EYE HOSPITAL, PONDICHERRY, INDIA

3.1 Introduction

In recent times, the number of people affected by Type-II diabetic mellitus is very high, and it is estimated by WHO that it may rise to more than 360 million in 2030 [1]. The effect of diabetes can be observed in almost all the organs in our human body, but the foremost affected one is the retina of the eye. If left untreated for a longer duration of time may lead to vision loss. Diabetic maculopathy (DM) is a state of diabetic retinopathy in which the retina gets damaged and oozes out a fluid rich in fat, which gets deposited within the retinal layers in the macula region. This fluid, known as diabetic macular edema (DME), when moves toward the fovea, the region of best visual acuity in the macula leads to distorted central vision [2]. Due to DM, the layers of the retina may get altered, and this could be easily visualized using spectral domain optical coherence tomography imaging (SDOCT). The progression of DM could be monitored accurately by using the thickness maps of retinal layers, option available in optical coherence tomography (OCT) machine, and hence treatment outcomes are also easily assessed [3–5]. Moreover, SDOCT plays a crucial role in early diagnosis and for disease follow-up in the case of DM.

3.2 Structural changes of DME

A typical normal retinal OCT image is shown in Fig. 3.1A. In DM condition structure of DME may take different forms as shown in Fig. 3.1B−J. This may mainly include

(i) Intraretinal fluid—fluid or edema accumulating within the retinal layers
(ii) Cystoid macular edema or simply cystoids is characterized by an increase in retinal layer thickness, intraretinal layer reflectivity, dimension reduction in fovea depression, alterations in the regular structured retinal layers, and also with clearly visible cystoid spaces; and further, this could be mild cystoid with cystoid spaces and horizontal diameter with less than 300 μm, or intermediate cystoid with horizontal diameter between 300 and 600 μm, or severe cystoid with cystoid spaces and horizontal diameter of greater than 600 μm,
(iii) Serous macula detachment, which is characterized by hyporeflective area under the detached retina and exactly over the retinal pigment epithelium
(iv) Exudates—bright yellowish shadow spots. These are protein oozing out of damaged blood vessels [6].

Ten structural changes of DME are shown in Fig. 3.1B−J. In each category, 40 images were obtained from Aravind Eye Hospital, Puducherry.

Convolutional neural networks are found to be successful in the detection of the patterns involved in DM [7−10].

To design a specific convolution neural network (CNN) exclusively for automatic DME changes identification, thousands of images in all categories are needed. As there is no database with all the structural changes of DME states as mentioned earlier in Fig. 3.1, the images obtained from Aravind Eye Hospital alone are used in this work. For a small dataset, transfer learning is an optimal way of implementing deep learning in CNNs [11,12]. In this transfer learning method, already pretrained CNNs are used for training and validation. In this work, we have trained and validated DME images using 10 pretrained CNNs, namely AlexNet, VGGNet 16, VGGNet 19, ResNet 50, ResNet 100, ResNet 152, ResNext, Inception v2, Inception v3, and PNasNet. The block diagram depicting the workflow is shown in Fig. 3.2. Totally 400 images are used with 40 images in each category.

The deep CNNs are trained and validated for categorizing different structural changes of DM. The next section discusses briefly the architectures of the various networks used followed by results and discussion.

3.3 Convolutional neural networks

3.3.1 AlexNet

AlexNet contains eight layers. This AlexNet won the Image Net challenge in 2012 [13−15]. The architecture is a simple one, which has serially arranged five convolution layers along with a fully connected layer as shown in Fig. 3.3.

Sl No	DME Class	Sample Image
A.	Normal macula	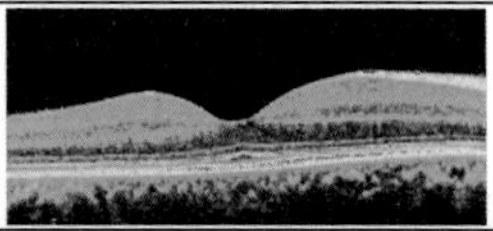
B.	Initial stage DM	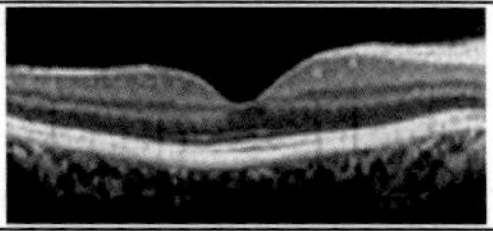
C.	cystoid	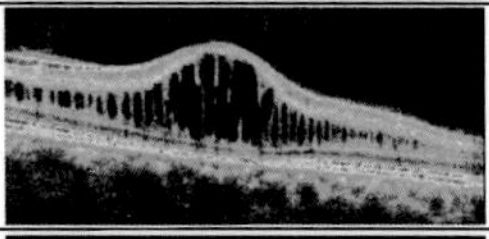
D.	Cystoid along with Intra Retinal Fluid (IRF)	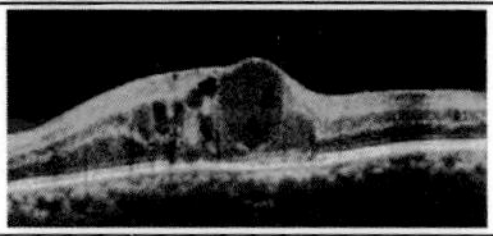
E.	Cystoid with IRF, exudates and SMD	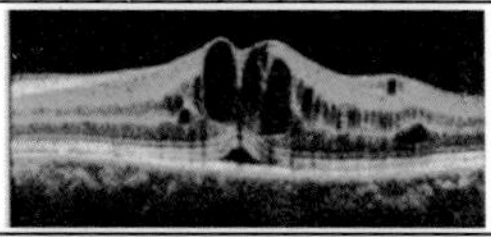
F.	Cystoid with exudates and Intra retinal fluid	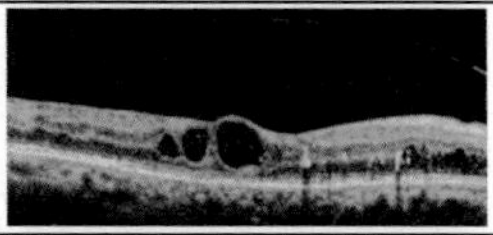
G.	Cystoid and serous macula detachment	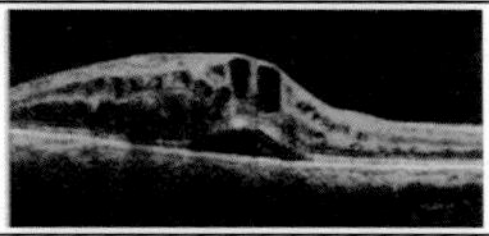
H.	Posterior hyaloidtraction(PHT) with IRF, cystoid and exudates.	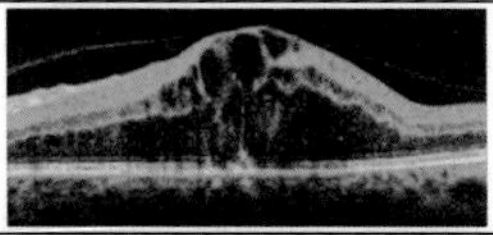
I.	Serous Macular Detachment (SMD)	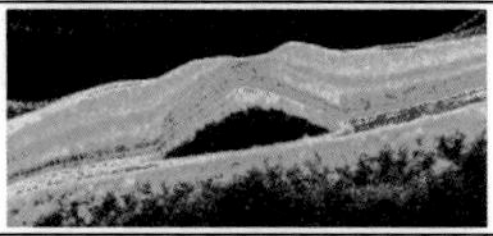
J.	SMD with IRF	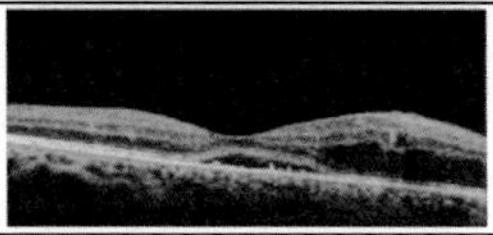

FIGURE 3.1 Structural changes of retinal layers in DME.

The first convolution layer has two groups of 48 kernels of size $11 \times 11 \times 3$ with a stride of 4 and zero padding. Output is $55 \times 55 \times 48$ feature maps of two groups. The second convolution layer has two groups of 128 kernels of size $5 \times 5 \times 48$ with a stride of 1 and padding of 2. Output is $27 \times 27 \times 128$ feature maps of two groups. Here, 3×3 overlapping max pooling of stride 2 is used [16]. The output is $13 \times 13 \times 128$ feature

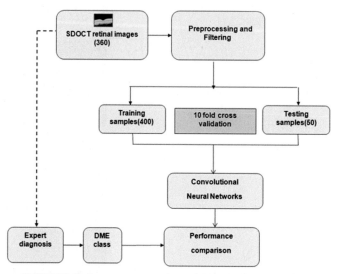

FIGURE 3.2 Performance comparison of CNNs in detection of DM.

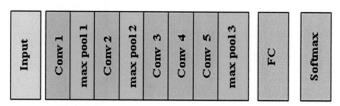

FIGURE 3.3 Architecture of AlexNet.

maps of two groups. The third convolution layer has two groups of 192 kernels of size $3 \times 3 \times 256$ with a stride of 1 and padding of 1. Output is $13 \times 13 \times 192$ feature maps of two groups, and 3×3 overlapping max pooling of stride 2 is used. The output is $13 \times 13 \times 128$ feature maps of two groups. Similar to the third convolution layer, the fourth convolution layer also has two groups of 192 kernels of size $3 \times 3 \times 256$ with a stride of 1 and padding of 1. The resultant output is two groups of feature maps of dimension $13 \times 13 \times 192$. Here a 3×3 overlapping max pooling of stride 2 is used. Again, the result is $13 \times 13 \times 128$ feature maps of 2 groups. The fifth convolution layer has 256 kernels of size $3 \times 3 \times 192$ with a stride of 1 and padding of 1. Outputs are $13 \times 13 \times 128$ feature maps of 2 groups. This layer is followed by a fully connected where all the neurons in the previous layer are connected to all the neurons in the fully connected layer. As mentioned earlier, in this work, the output has 10 categories; the fully connected layer has been designed for providing output in 10 nodes. In this AlexNet, totally there are around 60 million parameters that are to be trained. The network is validated through a confusion matrix, and the same arrived after the validation of images is shown in Fig. 3.4.

3.3.2   VGGNet

Visual Geometry Group (VGG) can be thought of as a successor of the AlexNet with more number of hidden layers, but it was created by a group named as VGG at Oxford's [17]. This VGG follows some of the techniques from its predecessors and improved on them. The layers in VGG 16 includes 13 convolutional layers and three fully connected layers and VGG 19 with 16 convolutional layer and three fully connected layers with max pool layer sandwiched between the convolutional layers [18]. The architecture of VGGNet is shown in Fig. 3.5.

Two major drawbacks with VGGNet: (1) It is painfully slow to train and (2) The network architecture weights themselves are quite large.

The confusion matrix for this network is shown in Fig. 3.6. For both 16-layer as well as 19- layer VGGNet, the same confusion matrix is arrived.

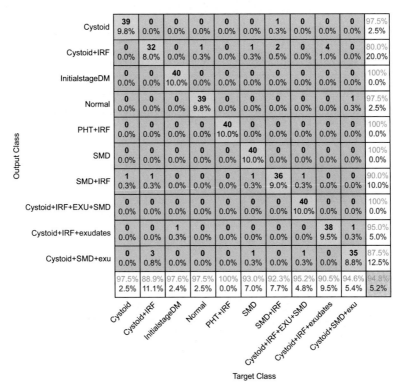

FIGURE 3.4 Confusion matrix of AlexNet.

FIGURE 3.5 Architecture of VGGNet.

Output Class / Target Class confusion matrix:

Output \ Target	Cystoid	Cystoid+IRF	InitialstageDM	Normal	PHT+IRF	SMD	SMD+IRF	Cystoid+IRF+EXU+SMD	Cystoid+IRF+exudates	Cystoid+SMD+exu	
Cystoid	40 10.0%	0 0.0%	0 0.0%	0 0.0%	0 0.0%	0 0.0%	0 0.0%	0 0.0%	0 0.0%	0 0.0%	100% 0.0%
Cystoid+IRF	0 0.0%	32 8.0%	0 0.0%	0 0.0%	0 0.0%	0 0.0%	3 0.8%	0 0.0%	1 0.3%	4 1.0%	80.0% 20.0%
InitialstageDM	0 0.0%	0 0.0%	40 10.0%	0 0.0%	0 0.0%	0 0.0%	0 0.0%	0 0.0%	0 0.0%	0 0.0%	100% 0.0%
Normal	2 0.5%	0 0.0%	0 0.0%	37 9.3%	0 0.0%	0 0.0%	0 0.0%	0 0.0%	0 0.0%	1 0.3%	92.5% 7.5%
PHT+IRF	0 0.0%	0 0.0%	0 0.0%	0 0.0%	40 10.0%	0 0.0%	0 0.0%	0 0.0%	0 0.0%	0 0.0%	100% 0.0%
SMD	0 0.0%	0 0.0%	0 0.0%	0 0.0%	0 0.0%	40 10.0%	0 0.0%	0 0.0%	0 0.0%	0 0.0%	100% 0.0%
SMD+IRF	0 0.0%	0 0.0%	0 0.0%	0 0.0%	0 0.0%	0 0.0%	39 9.8%	0 0.0%	0 0.0%	1 0.3%	97.5% 2.5%
Cystoid+IRF+EXU+SMD	0 0.0%	0 0.0%	0 0.0%	0 0.0%	0 0.0%	0 0.0%	0 0.0%	40 10.0%	0 0.0%	0 0.0%	100% 0.0%
Cystoid+IRF+exudates	1 0.3%	0 0.0%	0 0.0%	0 0.0%	0 0.0%	1 0.3%	0 0.0%	0 0.0%	38 9.5%	0 0.0%	95.0% 5.0%
Cystoid+SMD+exu	0 0.0%	0 0.0%	0 0.0%	0 0.0%	0 0.0%	0 0.0%	1 0.3%	0 0.0%	0 0.0%	39 9.8%	97.5% 2.5%
	93.0% 7.0%	100% 0.0%	100% 0.0%	100% 0.0%	100% 0.0%	97.6% 2.4%	90.7% 9.3%	100% 0.0%	97.4% 2.6%	86.7% 13.3%	96.3% 3.7%

FIGURE 3.6 Confusion matrix of VGGNet.

3.3.3 ResNET

The residual network, also known as ResNet, has five stages and two important blocks, namely convolution block and identity block. Both blocks have three convolution layers each. The ResNet-50 has more than 20 million trainable parameters. The concept of skip connection in the identity block was first introduced in ResNet as shown in Fig. 3.7. It allows an alternate path for the gradient to flow and hence overcomes the problem of vanishing gradient [19]. Moreover, this type of skip connections helps to learn an identity function that performs equally well in both higher as well as lower order layers.

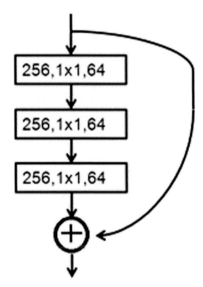

FIGURE 3.7 ResNet 50 Residual block.

The ResNet 50 (consisting of 50 layers), deep CNN architecture used for categorization of DME, is shown in Fig. 3.8 with the following features: (1) The purpose of the convolution layer is to combine serially the simpler patterns. For example, the features such as corners and edges are low-level features, which are used in the detection of primitive shapes and textures. (2) The residual network filters fetch more information in a more deep network [20]. These two reasons help to achieve good accuracy and very good performance by deep CNNs such as ResNets and finally the fully connected layer performs prediction by combining the detections of specific objects given by the convolution filters in the final layer. Also, average pooling and max pooling layers are included, which helps to integrate similar features into one and helps in dimensionality reduction of the extracted features. In the identity block, which is a more important unique feature of ResNet, the input activation and output activation have the same dimension. Finally, all the feature maps are integrated together by the fully connected layer and the final output is the value of the most probable class. For the reduction of execution time and to enhance the process a phenomenon known as batch normalization is also introduced. It is restrained to each mini-batch in the training process [19,21].

Similar to the above ResNet50, ResNet 100 and ResNet152 have 100 and 152 convolution layers, respectively. The ResNet confusion matrix is shown in Fig. 3.9. For all ResNets, we arrived at the same confusion matrix.

3.3.4   ResNeXt

The ResNeXt architecture shown in Fig. 3.10 can be visualized as an advancement of deep ResNet, in which the standard residual block is substituted by a split—transform

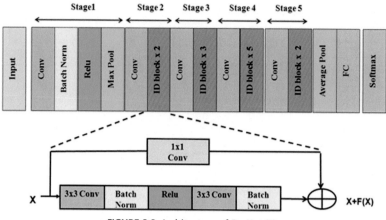

FIGURE 3.8 Architecture of ResNet 50.

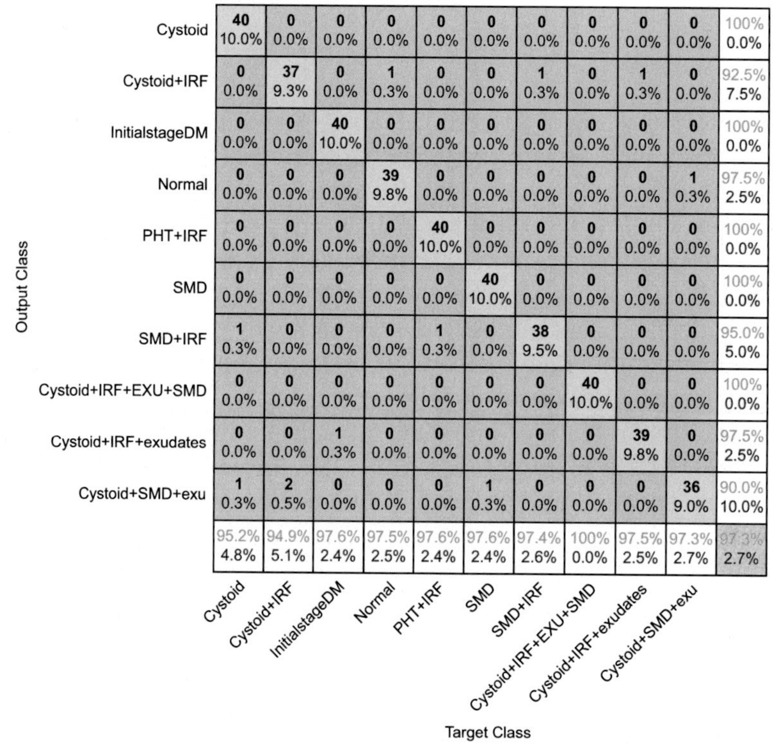

Output Class

	Cystoid	Cystoid+IRF	InitialstageDM	Normal	PHT+IRF	SMD	SMD+IRF	Cystoid+IRF+EXU+SMD	Cystoid+IRF+exudates	Cystoid+SMD+exu	
Cystoid	40 / 10.0%	0 / 0.0%	0 / 0.0%	0 / 0.0%	0 / 0.0%	0 / 0.0%	0 / 0.0%	0 / 0.0%	0 / 0.0%	0 / 0.0%	100% / 0.0%
Cystoid+IRF	0 / 0.0%	37 / 9.3%	0 / 0.0%	1 / 0.3%	0 / 0.0%	0 / 0.0%	1 / 0.3%	0 / 0.0%	1 / 0.3%	0 / 0.0%	92.5% / 7.5%
InitialstageDM	0 / 0.0%	0 / 0.0%	40 / 10.0%	0 / 0.0%	0 / 0.0%	0 / 0.0%	0 / 0.0%	0 / 0.0%	0 / 0.0%	0 / 0.0%	100% / 0.0%
Normal	0 / 0.0%	0 / 0.0%	0 / 0.0%	39 / 9.8%	0 / 0.0%	0 / 0.0%	0 / 0.0%	0 / 0.0%	0 / 0.0%	1 / 0.3%	97.5% / 2.5%
PHT+IRF	0 / 0.0%	0 / 0.0%	0 / 0.0%	0 / 0.0%	40 / 10.0%	0 / 0.0%	0 / 0.0%	0 / 0.0%	0 / 0.0%	0 / 0.0%	100% / 0.0%
SMD	0 / 0.0%	0 / 0.0%	0 / 0.0%	0 / 0.0%	0 / 0.0%	40 / 10.0%	0 / 0.0%	0 / 0.0%	0 / 0.0%	0 / 0.0%	100% / 0.0%
SMD+IRF	1 / 0.3%	0 / 0.0%	0 / 0.0%	0 / 0.0%	1 / 0.3%	0 / 0.0%	38 / 9.5%	0 / 0.0%	0 / 0.0%	0 / 0.0%	95.0% / 5.0%
Cystoid+IRF+EXU+SMD	0 / 0.0%	0 / 0.0%	0 / 0.0%	0 / 0.0%	0 / 0.0%	0 / 0.0%	0 / 0.0%	40 / 10.0%	0 / 0.0%	0 / 0.0%	100% / 0.0%
Cystoid+IRF+exudates	0 / 0.0%	0 / 0.0%	1 / 0.3%	0 / 0.0%	0 / 0.0%	0 / 0.0%	0 / 0.0%	0 / 0.0%	39 / 9.8%	0 / 0.0%	97.5% / 2.5%
Cystoid+SMD+exu	1 / 0.3%	2 / 0.5%	0 / 0.0%	0 / 0.0%	0 / 0.0%	1 / 0.3%	0 / 0.0%	0 / 0.0%	0 / 0.0%	36 / 9.0%	90.0% / 10.0%
	95.2% / 4.8%	94.9% / 5.1%	97.6% / 2.4%	97.5% / 2.5%	97.6% / 2.4%	97.6% / 2.4%	97.4% / 2.6%	100% / 0.0%	97.5% / 2.5%	97.3% / 2.7%	97.3% / 2.7%

Target Class

FIGURE 3.9 Confusion matrix of ResNet 50, ResNet 100, and ResNet152.

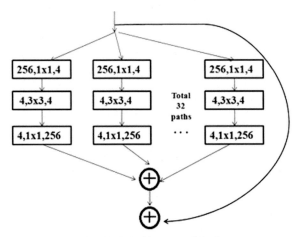

FIGURE 3.10 Architecture of ResNext

and merge strategy, similar to that of branched paths within a block implemented in the Inception models. Or in other words simply said, rather than convolutions performed over the entire input feature map, the block's input is anticipated into a number of smaller (channel) dimensional representations, each of which is subjected to a set of convolutional filters before being merged [22,23].

ResNeXt recycles a building block that combines a group of transformations with the same topology. In comparison to ResNet, it introduces a new dimension, cardinality (the size of the collection of transformations), as well as the depth and breadth dimensions.

The confusion matrix arrived at using this ResNext is shown in Fig. 3.11.

3.3.5   Inception v2

Inception makes use of parallel filters of different sizes and concatenates the output from them as shown in Fig. 3.12. The second generation of Inception network is v2 architecture, and it also makes use of the batch normalization feature [19].

Small changes in parameters are prevented from amplifying into larger and inefficient changes in activations in gradients by normalizing activations throughout the network; to be more specific, it prevents the training from becoming trapped in saturated nonlinearity regimes. Batch normalization also improves training's resistance to parameter scaling [24]. Due to the benefits of batch normalizing, other improvements include reducing dropouts and removing local response normalization. There were less

Output Class

	Cystoid	Cystoid+IRF	InitialstageDM	Normal	PHT+IRF	SMD	SMD+IRF	Cystoid+IRF+EXU+SMD	Cystoid+IRF+exudates	Cystoid+SMD+exu	
Cystoid	**40** 10.0%	0 0.0%	0 0.0%	0 0.0%	0 0.0%	0 0.0%	0 0.0%	0 0.0%	0 0.0%	0 0.0%	100% 0.0%
Cystoid+IRF	0 0.0%	**39** 9.8%	0 0.0%	0 0.0%	0 0.0%	0 0.0%	0 0.0%	0 0.0%	1 0.3%	0 0.0%	97.5% 2.5%
InitialstageDM	0 0.0%	0 0.0%	**40** 10.0%	0 0.0%	0 0.0%	0 0.0%	0 0.0%	0 0.0%	0 0.0%	0 0.0%	100% 0.0%
Normal	0 0.0%	0 0.0%	0 0.0%	**39** 9.8%	0 0.0%	0 0.0%	0 0.0%	0 0.0%	0 0.0%	1 0.3%	97.5% 2.5%
PHT+IRF	0 0.0%	0 0.0%	0 0.0%	0 0.0%	**40** 10.0%	0 0.0%	0 0.0%	0 0.0%	0 0.0%	0 0.0%	100% 0.0%
SMD	0 0.0%	0 0.0%	0 0.0%	0 0.0%	0 0.0%	**40** 10.0%	0 0.0%	0 0.0%	0 0.0%	0 0.0%	100% 0.0%
SMD+IRF	1 0.3%	0 0.0%	0 0.0%	0 0.0%	0 0.0%	0 0.0%	**39** 9.8%	0 0.0%	0 0.0%	0 0.0%	97.5% 2.5%
Cystoid+IRF+EXU+SMD	0 0.0%	0 0.0%	0 0.0%	0 0.0%	0 0.0%	0 0.0%	0 0.0%	**40** 10.0%	0 0.0%	0 0.0%	100% 0.0%
Cystoid+IRF+exudates	0 0.0%	1 0.3%	1 0.3%	0 0.0%	0 0.0%	0 0.0%	0 0.0%	0 0.0%	**38** 9.5%	0 0.0%	95.0% 5.0%
Cystoid+SMD+exu	0 0.0%	1 0.3%	0 0.0%	0 0.0%	0 0.0%	0 0.0%	0 0.0%	0 0.0%	0 0.0%	**39** 9.8%	97.5% 2.5%
	97.6% 2.4%	95.1% 4.9%	97.6% 2.4%	100% 0.0%	100% 0.0%	100% 0.0%	100% 0.0%	100% 0.0%	97.4% 2.6%	97.5% 2.5%	98.5% 1.5%

Target Class

FIGURE 3.11 Confusion matrix of ResNext

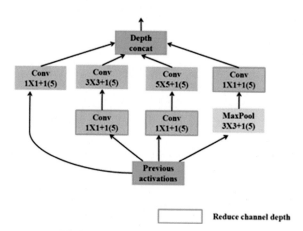

Reduce channel depth

FIGURE 3.12 Architecture of inception v2.

than 25 million parameters in this model. The researchers who developed this model factorized 5 × 5 convolutions into two stacked 3 × 3 convolutions that resulted in less computation cost while maintaining high quality. The confusion matrix for this Inception v2 is shown in Fig. 3.13.

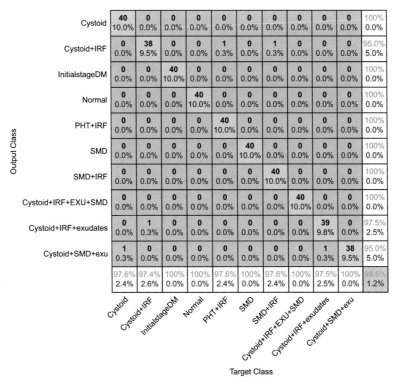

FIGURE 3.13 Confusion matrix of Inception v2.

3.3.6 Inception v3

Inception v3 is comparable to Inception v2 and has all of its features, with the exceptions listed in the following [25]. 5 × 5 convolutions in the previous model were replaced by two successive 3 × 3 convolutions as shown in Fig. 3.14, and the confusion matrix for the same is shown in Fig. 3.15.

3.3.7 PNASNet

This Progressive Neural Architecture Search Network (PNasNet) is a simple to complex approach network that has other strategies that directly search in the space of fully described structures with various advantages [26,27].

First, basic structures train faster, so we obtain some early results and can quickly train the surrogate. Second, we only ask the surrogate to estimate the quality of structures that are slightly different (bigger) than those it has already seen (trust-region methods). Third, we split the search space into a collection of smaller search spaces, potentially allowing us to explore models with many more blocks [28,29]. This network

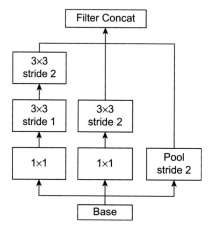

FIGURE 3.14 Architecture of Inception v3.

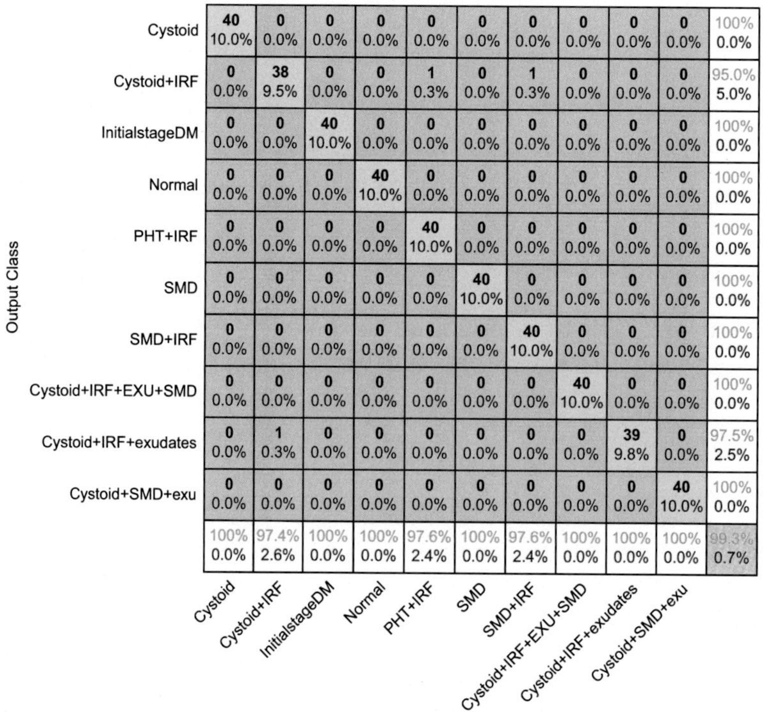

FIGURE 3.15 Confusion matrix of Inception v3.

makes use of reinforced learning. The architecture of PNasNet is shown in Fig. 3.16 and the confusion matrix is shown in Fig. 3.17.

3.4 Results and discussion

The networks, namely AlexNet, VGGNet 16, VGGNet 19, ResNet 50, ResNet 100, ResNet 152, ResNext, Inception v2, Inception v3, and PNASNet, are trained with 400 images, which include normal as well as all the 10 categories of DM structural changes and tested with 40 images as shown in Fig. 3.1. The validation accuracy is estimated for all the networks. Among all the networks, PNASNet performance is found to be excellent nearing 99.8% accuracy. As shown in Fig. 3.18, all the networks are compared in terms of the statistical parameters such as specificity, sensitivity, accuracy, precision, and F1 score.

The overall accuracy of all the networks is given in Table 3.1.

The overall accuracy of CNNs is given in Table 3.1. Although all the deep CNN performance is good in the classification of DM structural changes, it is observed that the overall accuracy of PNasNet is excellent. All the statistical parameters derived from the confusion matrix of PNasNet are given in Table 3.2.

Except for cystoids with IRF condition and Cystoid with SMD and exudates condition, all the other cases of structural changes have good validation accuracy, specificity, precision, and F1 score.

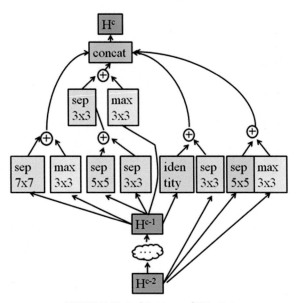

FIGURE 3.16 Architecture of PNasNet

Output Class

	Cystoid	Cystoid+IRF	InitialstageDM	Normal	PHT+IRF	SMD	SMD+IRF	Cystoid+IRF+EXU+SMD	Cystoid+IRF+exudates	Cystoid+SMD+exu	
Cystoid	40 10.0%	0 0.0%	0 0.0%	0 0.0%	0 0.0%	0 0.0%	0 0.0%	0 0.0%	0 0.0%	0 0.0%	100% 0.0%
Cystoid+IRF	0 0.0%	40 10.0%	0 0.0%	0 0.0%	0 0.0%	0 0.0%	0 0.0%	0 0.0%	0 0.0%	0 0.0%	100% 0.0%
InitialstageDM	0 0.0%	0 0.0%	40 10.0%	0 0.0%	0 0.0%	0 0.0%	0 0.0%	0 0.0%	0 0.0%	0 0.0%	100% 0.0%
Normal	0 0.0%	0 0.0%	0 0.0%	40 10.0%	0 0.0%	0 0.0%	0 0.0%	0 0.0%	0 0.0%	0 0.0%	100% 0.0%
PHT+IRF	0 0.0%	0 0.0%	0 0.0%	0 0.0%	40 10.0%	0 0.0%	0 0.0%	0 0.0%	0 0.0%	0 0.0%	100% 0.0%
SMD	0 0.0%	0 0.0%	0 0.0%	0 0.0%	0 0.0%	40 10.0%	0 0.0%	0 0.0%	0 0.0%	0 0.0%	100% 0.0%
SMD+IRF	0 0.0%	0 0.0%	0 0.0%	0 0.0%	0 0.0%	0 0.0%	40 10.0%	0 0.0%	0 0.0%	0 0.0%	100% 0.0%
Cystoid+IRF+EXU+SMD	0 0.0%	0 0.0%	0 0.0%	0 0.0%	0 0.0%	0 0.0%	0 0.0%	40 10.0%	0 0.0%	0 0.0%	100% 0.0%
Cystoid+IRF+exudates	0 0.0%	0 0.0%	0 0.0%	0 0.0%	0 0.0%	0 0.0%	0 0.0%	0 0.0%	40 10.0%	0 0.0%	100% 0.0%
Cystoid+SMD+exu	0 0.0%	1 0.3%	0 0.0%	0 0.0%	0 0.0%	0 0.0%	0 0.0%	0 0.0%	0 0.0%	39 9.8%	97.5% 2.5%
	100% 0.0%	97.6% 2.4%	100% 0.0%	100% 0.0%	100% 0.0%	100% 0.0%	100% 0.0%	100% 0.0%	100% 0.0%	100% 0.0%	99.8% 0.2%

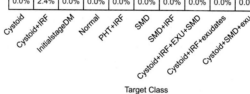

Target Class

FIGURE 3.17 Confusion matrix of PNasNet

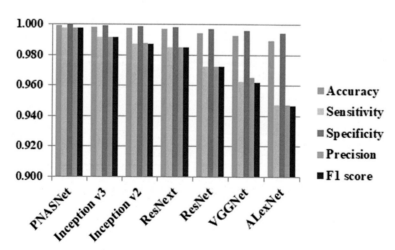

FIGURE 3.18 Comparison of CNNs in DM detection.

Table 3.1   Accuracy of CNNs.

Sl No	Type of CNN	Overall accuracy
1	AlexNet	94.8
2	VGGNet 16	96.3
3	VGGNet 19	96.3
4	ResNet-50	97.3
5	ResNet-100	97.3
6	ResNet-152	97.3
7	ResNeXt	98.5
8	Inception v2	98.8
9	Inception v3	99.3
10	PNasNet	99.8

Table 3.2   Statistical parameters of the PNASNet.

	Accuracy	Sensitivity	Specificity	Precision	F1 score
Cystoid	1.000	1.000	1.000	1.000	1.000
Cystoid with IRF	0.998	1.000	0.997	0.976	0.988
Initial stage DM	1.000	1.000	1.000	1.000	1.000
Normal	1.000	1.000	1.000	1.000	1.000
PHT with IRF	1.000	1.000	1.000	1.000	1.000
SMD	1.000	1.000	1.000	1.000	1.000
SMD with IRF	1.000	1.000	1.000	1.000	1.000
Cystoid and IRF and exudates and SMD	1.000	1.000	1.000	1.000	1.000
Cystoid and IRF and exudates	1.000	1.000	1.000	1.000	1.000
Cystoid and SMD and exudates	0.998	0.975	1.000	1.000	0.987

3.5 Conclusion

In this chapter, various deep CNN architectures for the detection of diabetic maculopathy are discussed in detail with their architecture and compared with validation results. The networks are designed to classify 10 various structural changes of DM including the Normal state. The structural changes include initial stage DM, cystoids, cystoid along with intraretinal fluid (IRF), cystoid with IRF, posterior hyaloid traction with IRF, serous macular detachment (SMD), SMD with IRF, cystoid, and IRF and exudates, and finally cystoids with SMD and exudates. A total of 400 DM images with 40 images in each category is taken into account. All the images were trained and validated using deep CNNs, namely AlexNet, VGGNet 16,VGGNet 19, ResNet 50, ResNet 100, ResNet152, ResNext, Inception v2, Inception v3, and PNasNet. Finally, their performance is compared in terms of statistical parameters derived through a confusion matrix. As there is no database existing on the net with all the structural categories mentioned earlier, the images obtained from Aravind Eye Hospital alone are used in this work. As the number of images is only limited, it is impossible to design and train a new deep CNN. Hence, transfer learning is implemented in the deep networks, and the best one among these is identified by comparing their performance in terms of accuracy, sensitivity, and specificity. It was observed that the overall accuracy of PNasNet is very much higher when compared with other networks. Also, as a future work, more images collected from hospitals from different imaging models will be taken into account and the statistical parameters will be compared.

References

[1] S. Wild, G. Roglic, A. Green, R. Sicree, H. King, Global prevalence of diabetes: estimates for the year 2000 and projections for 2030, Diabetes Care 27 (5) (2004) 1047−1053.

[2] G.H. Bresnick, Diabetic macular edema: a review, Ophthalmology 93 (7) (1986) 989−997.

[3] S.J. Chiu, X.T. Li, P. Nicholas, C.A. Toth, J.A. Izatt, S. Farsiu, Automatic segmentation of seven retinal layers in SDOCT images congruent with expert manual segmentation, Optics Express 18 (18) (2010) 19413−19428.

[4] A.G. Podoleanu, J.A. Rogers, D.A. Jackson, S. Dunne, Three dimensional OCT images from retina and skin, Optics Express 7 (9) (2000) 292−298.

[5] C. Framme, A. Walter, P. Prahs, R. Regler, D. Theisen-Kunde, C. Alt, R. Brinkmann, Structural changes of the retina after conventional laser photocoagulation and selective retina treatment (SRT) in spectral domain OCT, Current Eye Research 34 (7) (2009) 568−579.

[6] T. Otani, S. Kishi, Y. Maruyama, Patterns of diabetic macular edema with optical coherence tomography, American Journal of Ophthalmology 127 (6) (1999) 688−693.

[7] C.S. Lee, D.M. Baughman, A.Y. Lee, Deep learning is effective for classifying normal versus age-related macular degeneration OCT images, Ophthalmology Retina 1 (4) (2017) 322−327.

[8] F. Li, H. Chen, Z. Liu, Deep learning-based automated detection of retinal diseases using optical coherence tomography images, Biomedical Optics Express 10 (12) (2019) 6204−6226.

[9] M. Raghu, C. Zhang, J. Kleinberg, S. Bengio, Transfusion: Understanding transfer learning for medical imaging, Advances in Neural Information Processing Systems 32 (2019).

[10] D.S. Kermany, M. Goldbaum, W. Cai, C.C. Valentim, H. Liang, S.L. Baxter, J. Dong, Identifying medical diagnoses and treatable diseases by image-based deep learning, Cell 172 (5) (2018) 1122−1131.

[11] S.P.K. Karri, D. Chakraborty, J. Chatterjee, Transfer learning based classification of optical coherence tomography images with diabetic macular edema and dry age-related macular degeneration, Biomedical Optics Express 8 (2) (2017) 579−592.

[12] N. Padmasini, R. Umamaheswari, Automated detection of multiple structural changes of diabetic macular oedema in SDOCT retinal images through transfer learning in CNNs, IET Image Processing 14 (16) (2020) 4067−4075.

[13] A. Krizhevsky, I. Sutskever, G.E. Hinton, Imagenet classification with deep convolutional neural networks, Advances in Neural Information Processing Systems 25 (2012) 1097−1105.

[14] T. Shanthi, R.S. Sabeenian, Modified Alexnet architecture for classification of diabetic retinopathy images, Computers & Electrical Engineering 76 (2019) 56−64.

[15] Z. Jiang, H. Zhang, Y. Wang, S.-B. Ko, Retinal blood vessel segmentation using fully convolutional network with transfer learning, Computerized Medical Imaging and Graphics 68 (2018) 1−15.

[16] M.Z. Alom, T.M. Taha, C. Yakopcic, S. Westberg, P. Sidike, M.S. Nasrin, V.K. Asari, The History Began from Alexnet: A Comprehensive Survey on Deep Learning Approaches, 2018 arXiv preprint arXiv:1803.01164.

[17] K. Simonyan, A. Zisserman, Very Deep Convolutional Networks for Large-Scale Image Recognition, 2014 arXiv preprint arXiv:1409.1556.

[18] G. Altan, DeepOCT: an explainable deep learning architecture to analyze macular edema on OCT images, Engineering Science and Technology, an International Journal 34 (2022) 101091.

[19] C. Szegedy, S. Ioffe, V. Vanhoucke, A.A. Alemi, Inception-v4, inception-resnet and the impact of residual connections on learning, in: Thirty-first AAAI Conference on Artificial Intelligence, 2017.

[20] S. Targ, D. Almeida, K. Lyman, Resnet in Resnet: Generalizing Residual Architectures, 2016 arXiv preprint arXiv:1603.08029.

[21] D. McNeely-White, J.R. Beveridge, B.A. Draper, Inception and ResNet features are (almost) equivalent, Cognitive Systems Research 59 (2020) 312−318.

[22] R. Jain, M. Gupta, S. Taneja, D.J. Hemanth, Deep learning based detection and analysis of COVID-19 on chest X-ray images, Applied Intelligence 51 (3) (2021) 1690−1700.

[23] Y.P. Liu, Z. Li, C. Xu, J. Li, R. Liang, Referable diabetic retinopathy identification from eye fundus images with weighted path for convolutional neural network, Artificial Intelligence in Medicine 99 (2019) 101694.

[24] M. Dyrmann, H. Karstoft, H.S. Midtiby, Plant species classification using deep convolutional neural network, Biosystems Engineering 151 (2016) 72−80.

[25] R. Jain, P. Nagrath, G. Kataria, V.S. Kaushik, D.J. Hemanth, Pneumonia detection in chest X-ray images using convolutional neural networks and transfer learning, Measurement 165 (2020) 108046.

[26] C. Liu, B. Zoph, M. Neumann, J. Shlens, W. Hua, L.J. Li, K. Murphy, Progressive neural architecture search, in: Proceedings of the European Conference on Computer Vision (ECCV), 2018, pp. 19−34.

[27] M.A.A. Milton, Automated Skin Lesion Classification Using Ensemble of Deep Neural Networks in ISIC 2018: Skin Lesion Analysis towards Melanoma Detection Challenge, 2019 arXiv preprint arXiv: 1901.10802.

[28] S. Higa, Y. Iwashita, K. Otsu, M. Ono, O. Lamarre, A. Didier, M. Hoffmann, Vision-based estimation of driving energy for planetary rovers using deep learning and terramechanics, IEEE Robotics and Automation Letters 4 (4) (2019) 3876−3883.

[29] A. Khan, A. Sohail, U. Zahoora, A.S. Qureshi, A survey of the recent architectures of deep convolutional neural networks, Artificial Intelligence Review 53 (8) (2020) 5455−5516.

4

Epidemiological surveillance of blindness using deep learning approaches

Kurubaran Ganasegeran[1], Mohd Kamarulariffin Kamarudin[2]

[1]CLINICAL RESEARCH CENTER, SEBERANG JAYA HOSPITAL, MINISTRY OF HEALTH MALAYSIA SEBERANG PERAI, PENANG, MALAYSIA; [2]DEPARTMENT OF SOCIAL AND PREVENTIVE MEDICINE, FACULTY OF MEDICINE, UNIVERSITY OF MALAYA KUALA LUMPUR, MALAYSIA

4.1 Conceptualizing surveillance systems in ophthalmic epidemiology

Global public health headlines have been consistently hit with the rising burden of blindness in adults (Fig. 4.1), mostly afflicting communities from the low- and middle-income countries as a consequence of shifting demographics and population aging. Such a debilitating phenomenon could pose substantial dependence of an individual's daily living and impair quality of life by not only interrupting income or educational opportunities, but also escalating the risks of mortality [1]. But to the fact of highlighting the extent of the problem's magnitude, it is worthwhile to scrutinize that most blindness or visual impairment cases reported globally were predominantly classified as avoidable! Such circumstances are primarily observed in developing countries where national health surveillance systems are fundamentally weak to capture key health indicators through established disease or morbidity registers, deficits in health screening capacities or tools, and barriers to healthcare accessibility by communities. To further conceptualize the importance of surveillance systems, it would be crucial to appraise the case vignette as enumerated in Box Exhibit 4.1.

The scenario depicted in Box Exhibit 4.1 suggests that a large proportion of communities residing within the rural settings were at risk of being afflicted by cataracts. While public health authorities were able to make prompt decisions as a result of a community-based screening evaluation report, they could have acted earlier to institute appropriate health interventions to control the problem. But what prevented them from not doing so? The response to this question relies on the availability of a routine, systematic collection of data. Public health decisions solely rely on the availability of

GLOBAL PUBLIC HEALTH HEADLINES...

THE TIMES OF INDIA

'10,000 in Vadodara partially or fully blind owing to cataract'

Feb 6, 2022, 04.23 AM IST

Vadodara: As many as 10,612 persons are believed to have partial or complete blindness in the areas of Vadodara district outside city limits. Health officials have now taken up a door-to-door drive to treat these persons.

Under the National Programme for Control of Blindness and Visual Impairment, the health department conducts cataract operations and other activities round the year. The door-to-door visits will be taken up as a part of the programme. Vadodara district health officer Dr Ashok Jain said that it was believed that according to estimates there were 10,612 persons who were fully or partially blind due to cataract in the district. He said that this population will also be operated by May. The 80,606 operations conducted during the three years include 38,569 in 2019-20, 15,636 in 2020-21 and 26,401 in 2021-22.tnn

New data shows 33 million people are living with avoidable blindness and 260 million with avoidable moderate-to-severe visual impairment simply because they can't access the right kind of care.

Sci Dev Net

Bringing science & development together through news & analysis

05/10/17

Africa and Asia lead in proportion of blind adults

🕐 **Speed read**

- In 2015, 36 million people globally were blind, with most in Africa and Asia
- Researchers project the figure to about 115 million in 2050
- Countries should address blindness and eye other issues, experts urge

By: Stephanie Achieng'

Telangana Today

Sunday, February 6, 2022

Home   Hyderabad   Telangana   Andhra Pradesh   India   World   Entertainment   Science and Tech   Sport   Business   ...

Lifestyle › Glaucoma, a leading cause of irreversible blindness

Glaucoma, a leading cause of irreversible blindness

BY IANS   PUBLISHED: 1ST DEC 2021 9:35 AM

LATEST NEWS

- Kuch Dil Ne Kaha
- Gutta Jwala & Vishnu Vishal participates in Green India Challenge
- Golden era of music ended: Bollywood pays tribute to Lata Mangeshkar
- CM KCR condoles demise of Lata Mangeshkar

FIGURE 4.1 The headlines—global burden of vision loss or blindness.

BOX EXHIBIT 4.1 Community case vignette—cataract afflicted populations

A voluntary medical team had recently organized a medical camp in a rural district with approximately 8987 inhabitants within the coastal region of Country A in late February 2022. Inhabitants from three interconnected villages were approached, mostly over the age of 65 years, being inclined to their conventional agricultural, fishery, plantation, or forestry activities, and having no formal education. Their accessibility to healthcare facilities was limited. This 3-day medical screening program involved 1381 villagers undergoing basic medical and eye examinations. During the screening, the medical team found that approximately 39% of these villagers had visual acuity reduced to counting fingers, while around 19% had visual acuity reduced to hand movements or perception of light. Ophthalmoscopy examinations revealed that the fundus could not be detailed in approximately 10% of those with poor visual acuity. They concluded that the screening sample had a high prevalence of cataracts. Most of them had uncontrolled diabetes. The report was presented to the local health department and brought forward to the state and federal health departments for possible interventions. The response path to this problem was scrutinized as follows:

   I. What is the problem?
      High prevalence of cataracts among the villagers.
  II. What could have caused the problem?
      Probable risk factors that could have caused the high number of cataract cases were older aged people, low socioeconomic status, no formal education, uncontrolled diabetes, and poor accessibility to healthcare services.
 III. What could be done?
      - Free surgery, or
      - Free surgery plus reimbursement of transport costs, or
      - Free surgery plus free transport to and from the eye specialist's hospital
  IV. How should it be done?
      Federal policymakers have concluded that these populations being afflicted by cataracts would be given free surgery plus free transport to and from the specialist hospital. The state health department will provide the transport services to the nearest eye specialist hospital. The district health office was instructed to capture all inhabitants afflicted with cataracts. Public health policymakers have directed routine screenings to be implemented in the whole district, have health promotional and awareness campaigns, and provide mobile health clinics or transport to control the health problem.

comprehensive data collection to critically understand the distribution and determinants of health-related states within communities. It is because the medical team was able to capture a substantial number of cataract cases within communities that public health practitioners acted promptly to intervene! They actually followed the public health response path as exhibited in Fig. 4.2. But with such medical camps, it would not be feasible to provide comprehensive data for the whole community or

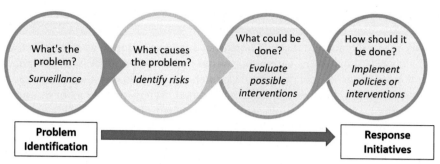

FIGURE 4.2 Public health response path.

population as it is cost-, time-, resource-, and labor-intensive. However, the availability of comprehensive data relies heavily on strong surveillance systems. The public health surveillance system is ideally conceptualized as an "ongoing, systematic collection, analysis, and interpretation of health-related data that are essential to the planning, implementation, and evaluation of public health practice, closely integrated with timely dissemination of the data to those responsible for prevention and control of diseases [2]." A useful surveillance system should direct appropriate policies or interventions to control health events. Surveillance systems may be categorized into:

I. **Active surveillance (sentinel surveillance)**—such a surveillance strategy identifies cases using community health screening surveys (as enumerated in Box Exhibit 4.1) or reviews of individual case records. They are substantially costly and labor-intensive, but of paramount use when cases have difficulty accessing healthcare services, or if there is an emerging disease that needs to be eliminated within a short amount of time that requires ongoing monitoring.

II. **Passive surveillance**—this approach detects cases when they seek healthcare services. Healthcare providers will routinely collect data on basic demographics, diagnoses, physical examination findings, or laboratory investigations conducted for patients to be captured within a centralized database system.

It would be essentially important for every country to build a functional surveillance system to initiate a response path for tackling blindness. A national-level sentinel surveillance system response is enumerated in Box Exhibit 4.2 to highlight the importance of such systems in tackling blindness.

The scenario appraised in Box Exhibit 4.2 shows the importance of capturing data on health-related states to identify the distribution of the disease, risk factors associated with the disease, outcomes from interventions implemented, and trends of the disease over time with or without interventions to control the health problem.

To date, developed countries were capacitated to maneuver disease registers for their country populations, but developing countries often struggle to maintain such registers at a national level. As a result, these countries opt for cross-sectional surveys that would

BOX EXHIBIT 4.2 Case vignette—national surveillance for tackling cataract in India

In line with the commensurate rise of global preventable blindness, India was heavily burdened by the relatively high incidence of cataract cases. The National Program for the Control of Blindness in India was launched in line with the World Bank Blindness Control Project (1994–2001) that fundamentally focused on eliminating cataract-related blindness in the country [3]. Thus, a National Sentinel Surveillance System was established through the Network of Sentinel Surveillance Units (SSUs) across different regions in India. This network of SSUs is supervised by the National Surveillance Unit, which would facilitate technical support to all 25 sentinel centers. The National Surveillance Unit (NSU) is supported by the World Health Organization to establish a framework that would strengthen information databases related to blindness.

The established sentinel sites throughout the country were equipped with appropriate infrastructures to capture and manage systemic data. The SSUs will submit quarterly reports regarding the number of cataract procedures performed and their end results, including postsurgical issues that arose to the NSU. The information tabulated in the database facilitated comprehensive assessments of the surgical procedures conducted, yet evaluators were able to determine the success rates stratified by sex, socioeconomic status, surgical procedures, the setting of the surgery performed, and subjects' age at the time of their surgery.

A national report on the surgical outcomes of five million cataract patients showed a rise in the proportion of cataract procedures accompanied by intraocular lens implantation (approximately 81% in 2001 as compared to 98% in 2011) and a reduction in surgeries among people with bilateral blindness (about 39% in 2001 in contrast to 17% in 2011). Surgical procedures were performed more in women than men. Visual acuity was better in the operated eyes after cataract surgeries. Postoperative complication rates were lower within a 10-year period (43.4/1000 surgeries in 2007 vs. 24.8/1000 surgeries in 2011). Common ocular morbidities reported in India were refractive errors (39.9%) and cataracts (18.8%) based on the 10 years' surveillance report.

I. What is the problem?

High incidence of cataracts in India.

II. What could have caused the problem?

The cases were prevalent in women and may be due to low socioeconomic status with difficult access to healthcare and surgical procedures for cataract removal.

III. What could be done?

Surgical procedures to remove the cataract and assess the outcomes over time.

IV. How should it be done?

The establishment of the National Surveillance Unit has enabled a systematic collection of data across all regions in India, and thus would enable policy advocates to comprehensively understand the distribution of cataract burden, their risks, and outcomes of postsurgical procedures.

capture health-related states at a particular point of time to provide a "snapshot" of blindness prevalence within a sample population to be studied. Although they are less costly, they are not generalizable or longitudinally powered to capture a representative population throughout the nation to understand the problem magnitude or establish the determinants of blindness over time. While several low-income countries opted for a survey-based approach or secondary data sources like census data to capture visual impairments or blindness prevalence across their populations in Nepal [4,5], Vanuatu [6], Paraguay [7], Chennai City [8], and Timor-Leste [9], these efforts were somewhat being able to highlight that the burden of blindness was mostly avoidable, yet specified risk factors such as socioeconomic determinants and accessibility to healthcare were highly associated with the prevalence of blindness or visual impairments in their studied populations. But they were not sufficiently being able to capture the trends over time or the effectiveness of interventions implemented by local health authorities, or neither they could be powered to describe the problem magnitude for country-level estimates stratified by different geographical levels or socioeconomic indicators attributed to blindness.

In recent times, efforts have been shifted toward utilizing passive surveillance, whereby established disease registers in local hospitals or healthcare facilities that capture sociodemographics, health profiles, diagnoses of blindness, interventions being conducted, and outcomes monitoring over time are being swiftly recorded in the database longitudinally at postfollow-ups of patients. For example, local registries managed by maxillo-facial surgeons and ophthalmologists in four Nigerian hospitals were able to capture blindness caused by facial trauma as a consequence of road traffic injuries, and subsequently be able to monitor postsurgical outcomes of affected patients [10]. While it was important to establish active surveillance in resource-limited settings, with the industrial revolution that has enabled economic inequity being narrowed down across different regions in a country, many governments worldwide are motivated to develop passive surveillance systems embedded within their respective healthcare systems. The collective accumulation of such disease registers may be centralized to an accessible data warehouse at a country level, yet a common database would enable health policymakers to capture a particular disease burden while assessing risk factors and outcomes of affected populations. Such efforts would enable policies to be implemented at the national level coherently to curb the burden of avoidable blindness. But with the rise of big health data and robust computational technologies, data analytics has been revolutionized toward the applications of artificial intelligence (AI), and the subset of AI known as deep learning algorithm has been widely applied to understand the burden of blindness and ophthalmic deficiencies efficiently [11]. Deep learning approaches were applied to detect blindness using retinal images [12] and predict the risk of diabetic retinopathy [13,14] recently.

A systematized literature review reported common visual impairments among older aged adults ($\geq$50 years) were cataract, glaucoma, age-related macular degeneration, and diabetic retinopathy [15]. With these emerging causes posing escalated risks for visual impairments and blindness, coupled with data revolution within healthcare that addressed needs gaps, this chapter would systematically appraise the importance of deep learning applications to diagnose, screen, and prevent avoidable blindness, facilitated by innovative real-time published examples from the literature.

4.2 Deep learning in ophthalmic epidemiological surveillance

Ophthalmology is a medical specialty that places a strong emphasis on visual cues. Optical coherence tomography (OCTs) images and digital color fundus photographs (CFPs) have been used in combination with visual fields to diagnose, monitor, and manage many retinal pathologies including diabetic retinopathy (DR), glaucoma, and age-related macular degeneration (AMD). The problem is that in typical settings, trained professionals are required to abstract information from these images, resulting in a disparity in diagnostic services in regions with limited human resources. It has been noted that more than half of the US counties have a shortage of ophthalmologists and optometrists per capita, and around 10% of people living with diabetes reside in counties where eye care experts are not available [16]. In developing countries, those figures are even more alarming.

Instead of relying on the presence of human experts, deep learning (DL) methods, for example, convolutional neural networks (CNNs) have provided an alternative way for diagnosing and segmenting the features of retinal disease. DL is a method that uses densely interconnected nodes to process information and perform tasks based on training examples similar to other artificial neural networks (ANNs). However, a DL classifier, for example, has a much greater number of hidden layers than ANNs.

CNN is widely used in image classification tasks because it was initially designed by simulating our understanding of human vision. Convolutional, pooling, and fully connected (FC) layers commonly make up CNN's hidden layers. Filter kernels in convolutional layers extract image properties like the texture and edges. Then, the FC layers assign a probability to the input image being a specific labeled class. A conventional image classification method would include a feature-engineering step in which specific visual features specified by experts are computed. This manual selection of features can lead to algorithms that are too specific, limiting their usefulness to other datasets. On the other hand, CNN avoids this by directly learning the most predictive features from images during weight training and optimization.

Using CNN-based models to detect retinal conditions from CFPs or OCTs has been shown in previous studies to be highly accurate. Therefore, it is possible to use them as a screening tool to quickly identify individuals who need additional attention and examination by eye specialists, to improve the diagnostic accuracy and speed. Fig. 4.3 depicts other possible uses of DL in epidemiological surveillance.

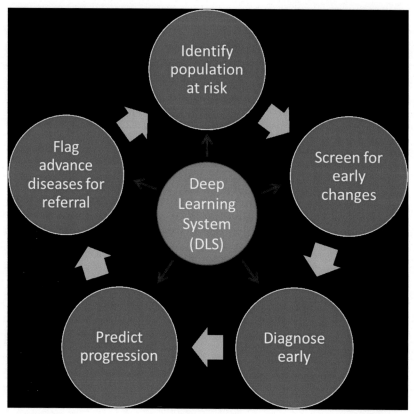

FIGURE 4.3 Uses of deep learning system (DLS) in epidemiological surveillance.

4.2.1 Application of DL in primary and secondary prevention

The term "primary prevention" refers to measures taken before the onset of disease to avert illness. Primary prevention in the fight against blindness, for example, entails identifying those at high risk of contracting the associated eye diseases. Secondary prevention focuses on the early detection and treatment of illnesses. The incidence and progression of vision loss can be significantly reduced and even reversed if detected and treated early. Many cases of visual loss in the community begin with asymptomatic conditions that gradually worsen. Thus, the strategy for preventing blindness must include all of the essential elements of primary and secondary prevention in epidemiological surveillance. These include identifying risk factors, screening and early detection of diseases, and early referrals for treatment.

To avoid late complications, at-risk patients should be identified as soon as possible and monitored at the appropriate intervals. Unfortunately, there is still an unmet need for all high-risk patients to have their retinas examined. Some remote communities cannot afford to see an ophthalmologist on a regular basis due to inequality in the distribution of health services. Thus, in underserved areas with limited access to healthcare, robust automated tools that evaluate fundus photographs that objectively predict disease progression could be a new way to overcome the limitations of resources. A DL system was found to be similar to ophthalmologists in classifying optic disc appearance from fundus examinations in a systematic review that included findings from various study settings [17]. Many studies have shown that as the disease progresses, DL algorithms may reliably anticipate its progression even before the onset of symptoms and detect the disease at an early stage with high sensitivity and specificity [11,14]. DL-based disease surveillance could therefore reduce the reliance and burden on manual graders significantly, lowering costs and increasing overall efficiency.

4.2.2 Application of DL in diabetic retinopathy surveillance

One of the most prevalent complications of diabetes is DR. Its diagnosis relies upon the detection of microvascular lesions as well as inflammation and neurodegeneration in the retina. The severity of DR is diagnosed through visual assessment of the fundus either by direct examination or evaluation of CFPs. Trained personnel will conduct the evaluation in accordance to standard criteria as described either in the Early Treatment of Diabetic Retinopathy Study protocol, the UK national guidelines on screening for DR, or the International Clinical Diabetic Retinopathy and Diabetic Macular Edema Severity Scale. By using these criteria, the severity of DR can be categorized into either none, mild, moderate, severe, or proliferative.

The primary goal of a DR retinal examination in a community setting is to refer patients who require additional evaluation by ophthalmologists promptly. A comprehensive assessment of DR-specific lesions and their severity is therefore not required during screening. Thus, many DL algorithms from prior research on DR assessment simplified DR severity level classification into either referable or nonreferable diseases [17]. Some studies have also used classes corresponding to no retinopathy versus DR of any severity. In those studies, moderate or severe DR is referred to as "referable DR" (RDR). This definition includes referable diabetic macular edema (DMO). Referable DMO is defined as the presence of hard exudates in the vicinity of the macula.

Automated classification can improve patient outcomes by increasing screening program efficiency and coverage, increasing access, and enabling early detection and treatment. Furthermore, it has been shown that this data-driven approach to DR screening is highly effective in various settings, resulting in a reliable decision about who requires further investigation. The following case study [Box Exhibit 4.3] examines how a DL model was constructed to classify populations at risk of DR with commendable predictive performance compared with eye specialists.

In addition to Ting et al. [18] numerous organizations have developed a broad spectrum of commercially accessible software for the AI-supported detection of DR, dubbed automated retinal image analysis systems (ARIAS). The research was undertaken by Tufail et al. [19] to assess some of these systems. Three ARIAS; iGradingM (UK),

BOX EXHIBIT 4.3 Deep learning for diabetic retinopathy

Introduction and objectives

Several AI-based algorithms have been able to accurately identify DR. However, clinical validation and implementation of the DL models in practice can be challenging. For the most part, researchers used datasets that were either collected from a homogenous sample of the population or were openly accessible to the public. This could lead to a model that is overly tailored to a specific dataset and lacks the ability to be generalized. Ting et al. [13] from Singapore used retinal images from a multiethnic cohort of people with diabetes to conduct their research. Their objective was to use those ethnically diverse images to train and validate a DL system (DLS) to identify RDR, vision-threatening DR, and other associated retinal pathologies (referable possible glaucoma and referable AMD).

Methods and analytical concepts
Design, settings, and participants

In this cross-sectional study, researchers collected retinal images from individuals with diabetes who participated in the Singapore National Diabetic Retinopathy Screening Program. Model training was done with images taken between 2010 and 2013.

Network architecture

The DLS is made up of eight CNNs that use a VGGNet architecture adaption:

(i). an ensemble of two networks for the classification of DR severity. One network was provided with the original images as input, while the other network was provided locally contrast-normalized images;

(ii). an ensemble of two networks for the identification of referable possible glaucoma;

(iii). an ensemble of two networks for the identification of referable AMD;

(iv). one network to assess image quality; and

(v). one network to reject invalid nonretinal images.

Preprocessing

Each retinal picture is first automatically segmented to retrieve only the retina disc. This region is then evenly rescaled to fit a 512×512 pixel standardized square template. Finally, a limited set of transformations to the input images was used to expose the convolutional network for additional plausible input feature variations, including mirroring, rotation, and scaling by a modest amount.

Model validation

Each network was trained to converge its performance against a small held-out validation set. External validation of RDR classification networks was also carried out against 10 additional multiethnic datasets from various countries, each having a distinct diabetes population.

BOX EXHIBIT 4.3   Deep learning for diabetic retinopathy—cont'd

Main outcomes and measures

Each of the DR network's output nodes was indexed from 0 to 4 according to the severity of the DR class. The mean of the outputs of the two convolutional networks was then used to get the final DR severity score. The test image outputs were classified using a threshold of 0.70 to differentiate between RDR and non-RDR as well as to detect vision-threatening DR (severe nonproliferative DR and proliferative DR).

The area under the receiver operating characteristic curve (AUC), sensitivity and specificity of the DLS were also compared to those of professionally qualified graders (retinal specialists, general ophthalmologists, or optometrists).

Results

For RDR, the AUC of the DLS was 0.936, and for vision-threatening DR, it was 0.958. The DLS exhibited similar sensitivity to trained graders in diagnosing RDR (90.5% vs. 91.1%; P-value = 0.68), but the professional graders had a greater specificity (91.6% vs. 99.3%; P-value <0.001). The DLS demonstrated better sensitivity than trained graders for vision-threatening DR (100% vs. 88.5%; P-value <0.001) but poorer specificity (91.1% vs. 99.6%; P-value <0.001).

Conclusion

In community-based DR screening research, a validated DL system has found good sensitivity and specificity for detecting RDR. This demonstrates the potential for AI technology to be used and adopted in a low-resource situation to lower the prevalence of avoidable blindness.

Retmarker (Portugal), and EyeArt (USA) were used to compare the automated versus conventional DR grading by human graders or ophthalmologists. RDR detection sensitivity was stronger in EyeArt and Retmarker than in human graders. Solanki et al. [20] employed the Messidor2 dataset, a publicly available dataset of DR, with their EyeArt AI engine. EyeArt's RDR screening sensitivity was 93.8%, the specificity was 72.2%, and the AUC was 0.94.

4.2.3   Application of deep learning in glaucoma surveillance

Glaucoma is a set of optic neuropathies that can result in vision loss associated with levels of intraocular pressure and progressive retinal ganglion cell destruction. Glaucoma develops in stages, beginning with asymptomatic preperimetric stages and progressing to functional impairment in the perimetric stages. Glaucomatous structural damage can be evaluated in detail with spectral-domain OCT (SD-OCT). This instrument measures the thickness of the macular ganglion cell complex and the retinal nerve fiber layer (RNFL). Glaucoma-induced retinal ganglion cell destruction can be gauged indirectly by the thickness of the RNFL as determined by the SD-OCT scans. Glaucomatous damage to these layers usually occurs in characteristic patterns, with considerable evidence showing the association between RNFL thickness loss and glaucoma.

DL algorithms for glaucoma detection are typically trained to classify and categorize the presence of glaucoma directly from the SD-OCT or optic disc photographs. However, some algorithms quantify glaucomatous structural changes by predicting the RNFL thickness from the assessment of the optic disc images [21]. A standard RNFL thickness normative database can then be used to categorize the optical disc images as normal or abnormal.

Many studies have shown that a DLS can detect preperimetric glaucoma even when standard automated perimetry detects no visual field abnormalities. The early detection of glaucomatous morphologic changes made possible by employing a sensitive screening technique is critical to preventing further vision loss. Some methodological highlights related to the use of DLS in glaucoma detection are provided in the following case study [Box Exhibit 4.4].

BOX EXHIBIT 4.4 Deep learning for glaucoma

Introduction and objectives

SD-OCT measurement of RNFL thickness necessitates precise delineation of the RNFL's anterior and posterior borders. If the RNFL is incorrectly segmented, glaucomatous damage cannot be accurately detected. However, it is possible to develop algorithms to detect glaucomatous changes without segmentation. Thompson et al. [22] developed DL algorithms to identify glaucomatous damage in SD-OCT images without the need for segmentation and compared that to the RNFL thickness parameters from the SD-OCT software.

Methods and analytical concepts
Design, setting, and participants

SD-OCT scans of glaucomatous eyes in both perimetric and preperimetric stages as well as normal eyes from Duke Glaucoma Repository were used in this cross-sectional study. At the patient level, the data from March 2008 to April 2019 set were separated into three categories: training (50%), validation (20%), and test (30%).

Network architecture

A residual deep CNN (ResNet34) architecture that was trained previously on the ImageNet dataset was utilized. The network was then fine-tuned with differential learning rates and Adam optimizer beginning with the last two layers and subsequently all layers by unfreezing them in sequence.

Preprocessing

The SD-OCT B-scans without segmentation lines were down-sampled to 496 × 496 pixels, then the pixel values were scaled to a range of 0–1. The images utilized did not have any segmentation lines; thus, the DL algorithm could figure out which features were important for predicting the existence of glaucoma independent of the pretraining segmentation.

BOX EXHIBIT 4.4   Deep learning for glaucoma—cont'd

    Random lighting adjustments of picture balance and contrast of up to 5%, horizontal image flips, and image rotations of up to 10° were also used to increase the variability of the training images. Transformations like flipping, translation, shearing, and rotation are frequently used in OCT imaging because they represent the actual variance that can be captured.

Model validation

The network performance was fine-tuned and tested against a validation and test set, respectively.

Main outcomes and measures

The network was built to discriminate between diseased and normal eyes and to provide an estimate of glaucoma likelihood. The DL algorithm's output was compared to the RNFL thickness provided by SD-OCT software. Network performances were evaluated by comparing the AUC and sensitivity at 80% and 95% specificity.

Results

To discriminate glaucoma from the healthy eye, the network exhibited considerably higher discrimination than the SD-OCT RNFL thickness (AUC 0.96 vs. 0.87) for each thickness sector (all P-value < 0.001). Sensitivities were estimated at fixed specificities of 80% and 95%. The DL system (81%; 95 CI, 64%−97%) has higher sensitivity than RNFL thickness at 95% specificity (67%; 95 CI, 58%−76%).

Conclusion

For identifying glaucomatous damage on OCT images, a DL algorithm without segmentation predicts better than standard RNFL thickness measures, particularly in early disease.

4.2.4   Deep learning in age-related macular degeneration surveillance

AMD may be divided into three stages: early, intermediate, and advanced based on the degree of retinal degeneration it manifests in the macula. The presence of drusen distinguishes the intermediate stage from the early stage, and both are usually asymptomatic. The macula can be damaged in either a "wet" or a "dry" variant of advanced AMD. The ingrowth of choroidal neovascularization (CNV) and fibrocytes characterizes neovascular AMD. Geographic atrophy (GA) of the retinal pigment epithelium (RPE) overlaying drusen in the central macula characterizes the advanced "dry" type of AMD.

    While some people with AMD may go years without symptoms due to the gradual course of the illness, others rapidly advance to one or both forms of late AMD [23]. Identifying those with the greatest risk of developing advanced AMD is therefore critical to begin treatment before irreversible damage occurs. Traditionally, CFPs were considered the gold standard for identifying early AMD. However, OCT has largely replaced

CFPs as a diagnostic tool in recent years because the former can offer three-dimensional cross-sectional anatomic information on retinal defects that CFPs cannot.

AMD epidemiological monitoring can benefit from AI-automated systems. In the case study that follows [Box Exhibit 4.5], we will see how researchers use DL to create a model for identifying people at risk of developing AMD.

BOX EXHIBIT 4.5 Deep learning for AMD

Introduction and objectives

AMD is a condition that affects millions of people, but the risk of developing into severe AMD and becoming legally blind varies greatly. Machine learning techniques applied to imaging data have shown promise in predicting AMD progression. By combining multiple imaging, demographic, and genetic features, Schmidt-Erfurth et al. [24] proposed machine learning approaches to predict individual conversion from intermediate to advanced AMD using OCT imaging.

Methods and analytical concepts

Design, setting, and participants

This was a study based on longitudinal data of participants in the HARBOR clinical trial who had neovascular AMD. HARBOR was a 24-month, phase III, multicenter, double-masked, active treatment-controlled research that looked at the efficacy and safety of intravitreal ranibizumab in patients with subfoveal neovascular AMD who had not previously received treatment. The research procedure requires that participants have monthly SD-OCT imaging exams for both eyes. The genotypes, as well as participants' demographic data, were identified. The genetic features were recorded as the number of risk alleles of single-nucleotide polymorphisms at 34 AMD-associated loci, whereas demographic factors were recorded for age, gender, and smoking status.

Data preprocessing and model development

A series of automated image analyses were done utilizing OCT analysis software and a CNN with an encoder-decoder architecture to extract OCT characteristics related to AMD. In those analyses, the retinal layer (outer neurosensory layers and RPE), drusen and reticular pseudodrusen, and hyperreflective foci were volumetrically segmented and shown as two-dimensional en face thickness maps. These maps yielded a variety of parameters related to AMD, such as the drusen number, total volume and area, and maximum, mean, and variability in height.

A sparse Cox proportional hazards (CPH) model was constructed employing the given quantitative features. The CPH model is a multivariable linear model for time-to-event data, in this study's context, the time when intermediate AMD becomes advanced AMD. Two separate predictive models: one to estimate the risk of conversion to CNV and the other to estimate the risk of conversion to GA were built. The least absolute shrinkage and selection operator was employed to regularize the model, which reduces the influence of the model predictors for a more generalizable forecast.

BOX EXHIBIT 4.5   Deep learning for AMD—cont'd

Model validation

The model performance was evaluated using a 10-fold cross-validation method, in which method, the dataset was split into 10 equal subsets (folds). One-fold (10% of the eyes) was kept as a test set, while the other nine folds (90% of the eyes) were utilized for model training.

Main outcomes and measures

The prediction models yielded a hazard ratio, which was used as an indicator of AMD conversion risk. The value was then utilized to create the receiver operating characteristics curve, which was used to assess the prediction model's ability to distinguish between converting and nonconverting eyes.

Results

Within 2 years, 159 eyes (32%) had progressed to advanced AMD, 114 to CNV, and 45 to GA. The prediction model was able to differentiate between converting and nonconverting eyes with an AUC of 0.68 for CNV and 0.80 for GA, respectively.

Conclusion

Personalized AMD progression prediction is possible with machine learning and automated imaging feature analysis.

4.3 Limitations

To build a reliable model and allow for more precise parameter training, a neural network requires a vast quantity of training data, necessitating the linking of multiple databases. Therefore, while developing a DLS, data access and sharing are crucial aspects to address upon. Open access to retinal images is especially an essential topic to discuss, but it is also a legal minefield. The issue of privacy protection is delicate in retinal imaging since anonymization is difficult to achieve due to the retinal vasculature architecture, which is unique for individual identification. As a result, these datasets are hard to come by.

Aside from being rare, existing datasets frequently contain only a small number of scans with pathological characteristics or only include images from a single OCT vendor. This is especially true when the datasets were obtained through national screening programs. Thus, data augmentation for the diseased cohort is often done simultaneously as sampling normal images to level the number of diseased and normal images in a study.

Many prior studies can be used as a guide for implementing a DLS. However, there are many different research methods, algorithms, and terminologies, making it difficult to make sense of the many different findings that come from these studies. Nonstandard imaging or postprocessing techniques among devices and inconsistent reporting metrices are just two examples of these problems [25].

It is possible that physicians may be reluctant to use a DLS in a clinical setting because a DLS, being a black box model, calculates the severity of retinal diseases based on the pixel intensities in a fundus picture. In addition, clinical data is rarely used in studies involving deep learning algorithms, making it hard to figure out how a neural network arrives at a given choice or which features it uses in deep learning. Future research should look at how the DL approach influences diagnostic judgments when paired with other clinical data like risk factors.

4.4 Conclusion

The prevention of blindness and eye care delivery should be integrated with general health services, especially primary care. However, program implementation should be based on available technology and appropriate resources for the country or region. The basis of epidemiological surveillance is that it should be developed and employed to produce accurate information to decision-makers promptly and in a cost-effective way. Therefore, the analytics methods must include techniques that would increase efficiency and close the gaps in service delivery, especially in resource-lacking areas. It has been shown that a well-developed DL system could replace trained professionals with a more cost-effective retinal condition screening and epidemiological surveillance.

The efficiency of surveillance might be further improved by personalizing data collection based on the individual likelihood of the development or progression of visual loss due to retinal diseases. Often, it is uncertain which patients fall into the fast versus slow progressor groups. A DL algorithm may produce this form of individual prediction. The traditional risk estimation model can only approximate the progression risk for groups of patients with comparable signs and symptoms and cannot precisely forecast the course of retinal diseases in an individual patient.

A DL-based automated tool for assessing retinal diseases could also provide a novel solution to reduce misdiagnosis while increasing accessibility to healthcare and promoting earlier treatments. It also reduces or potentially eliminates unnecessary referrals caused by the unavailability of trained personnel to assess and monitor retinal conditions. Thus, a DLS should become part of the comprehensive strategy that supports the continuous process of screening, monitoring, and referring of retinal diseases to have a significant impact on the public healthcare system. However, because false positives and false negatives might be exaggerated in settings with a large general population, a DL-based eye evaluation should only be used in conjunction with subsequent manual inspection when positive cases are detected. This will allow for more screening episodes at a cheaper cost while ensuring the same health outcomes.

Acknowledgment

We thank the Ministry of Health Malaysia for the support to publish this chapter.

Conflict of Interests

None declared.

Funding

None.

References

[1] C.A. McCarty, M.B. Nanjan, H.R. Taylor, Vision impairment predicts 5-year mortality, British Journal of Ophthalmology 85 (3) (March 2021) 322–326.

[2] S.B. Thacker, G.S. Birkhead, Surveillance, in: M.B. Gregg (Ed.), Field Epidemiology, Oxford University Press, Oxford, England, 2008.

[3] P. Vashist, N. Gupta, A. Rathore, Sentinel surveillance of blindness: an initiative of the national programme for control of blindness in India, Indian Journal of Community Medicine 37 (3) (Jul-Sep 2012) 139–141.

[4] L.B. Brilliant, R.P. Pokhrel, N.C. Grasset, J.M. Lepkowski, A. Kolstad, W. Hawks, R. Pararajasegaram, G.E. Brilliant, S. Gilbert, S.R. Shrestha, J. Kuo, Epidemiology of blindness in Nepal, Bulletin of the World Health Organization 63 (2) (1985) 375–386.

[5] R. Thapa, S. Bajimaya, G. Paudyal, S. Khanal, S. Tan, S.S. Thapa, G.H.M.B. van Rens, Prevalence and causes of low vision and blindness in an elderly population in Nepal: the Bhaktapur retina study, BMC Ophthalmology 18 (1) (February 2018) 42.

[6] H.S. Newland, M.F. Harris, M. Walland, D. McKnight, J.E. Galbraith, W. Iwasaki, K. Momomura, Epidemiology of blindness and visual impairment in Vanuatu, Bulletin of the World Health Organization 70 (3) (1992) 369–372.

[7] F. Yaacov-Pena, D. Jure, J. Ocampos, M. Samudio, J.M. Furtado, M. Carter, V.C. Lansingh, Prevalence and causes of blindness in an urban area of Paraguay, Arquivos Brasileiros de Oftalmologia 75 (5) (October 2012) 341–343.

[8] L. Vijaya, R. George, R. Asokan, L. Velumuri, S.V. Ramesh, Prevalence and causes of low vision and blindness in an urban population: the Chennai Glaucoma Study, Indian Journal of Ophthalmology 62 (4) (April 2014) 477–481.

[9] M. Correia, T. Das, J. Magno, B.M. Pereira, V. Andrade, H. Limburg, J. Trevelyan, J. Keeffe, N. Verma, Y. Sapkota, Prevalence and causes of blindness, visual impairment, and cataract surgery in Timor-Leste, Clinical Ophthalmology 11 (November 2017) 2125–2131.

[10] B.A. Famurewa, F.O. Oginni, B.A. Adewara, B. Fomete, C. Aniagor, B. Aluko-Olokun, R.E. Morgan, M.I. Amedari, Epidemiology of blindness from facial Trauma: a multi-centre Nigerian study, Craniomaxillofacial Trauma and Reconstruction (December 2021). https://doi.org/10.1177/19433875211060931 (Online first).

[11] Y. Tong, W. Lu, Y. Yu, Y. Shen, Application of machine learning in ophthalmic imaging modalities, Eye and Vision (Lond) 7 (22) (April 2020).

[12] A.K. Jaiswal, P. Tiwari, S. Kumar, M.S. Al-Rakhami, M. Alrashoud, A. Ghoneim, Deep learning-based smart IoT health system for blindness detection using retina images, IEEE Access 9 (May 2021) 70606–70615.

[13] D.S.W. Ting, C.Y. Cheung, Q. Nguyen, C. Sabanayagam, G. Lim, Z.W. Lim, G.S.W. Tan, Y.Q. Soh, L. Schmetterer, Y.X. Wang, J.B. Jonas, R. Varma, M.L. Lee, W. Hsu, E. Lamoureux, C.Y. Cheng, T.Y. Wong, Deep learning in estimating prevalence and systemic risk factors for diabetic retinopathy: a multi-ethnic study, NPJ Digital Medicine 2 (24) (April 2019).

[14] A. Bora, S. Balasubramanian, B. Babenko, S. Virmani, S. Venugopalan, A. Mitani, G. de Oliveira Marinho, J. Cuadros, P. Ruamviboonsuk, G.S. Corrado, L. Peng, D.R. Webster, A.V. Varadarajan, N. Hammel, Y. Liu, P. Bavishi, Predicting the risk of developing diabetic retinopathy using deep learning, The Lancet Digital Health 3 (1) (January 2021) e10–e19.

[15] GBD 2019 Blindness and Vision Impairment Collaborators; Vision Loss Expert Group of the Global Burden of Disease Study, Causes of blindness and vision impairment in 2020 and trends over 30 years, and prevalence of avoidable blindness in relation to VISION 2020: the Right to Sight: an analysis for the Global Burden of Disease Study, Lancet Global Health 9 (2) (February 2021) e144–e160. Erratum in: The Lancet Global Health, vol.9, no.4, p.e408, Apr 2021.

[16] R. Pires, S. Avila, J. Wainer, E. Valle, M.D. Abramoff, A. Rocha, A data-driven approach to referable diabetic retinopathy detection, Artificial Intelligence in Medicine 96 (May 2019) 93–106.

[17] M.M. Islam, H.C. Yang, T.N. Poly, W.S. Jian, Y.C. Li, Deep learning algorithms for detection of diabetic retinopathy in retinal fundus photographs: a systematic review and meta-analysis, Computer Methods and Programs in Biomedicine 191 (105320) (July 2020).

[18] D.S. Ting, C.Y. Cheung, G. Lim, G.S. Tan, N.D. Quang, A. Gan, H. Hamzah, R. Garcia-Franco, I.Y. San Yeo, S.Y. Lee, E.Y. Wong, Development and validation of a deep learning system for diabetic retinopathy and related eye diseases using retinal images from multiethnic populations with diabetes, JAMA 318 (22) (December 2017) 2211–2223.

[19] A. Tufail, C. Rudisill, C. Egan, V.V. Kapetanakis, S. Salas-Vega, C.G. Owen, A. Lee, V. Louw, J. Anderson, G. Liew, L. Bolter, Automated diabetic retinopathy image assessment software: diagnostic accuracy and cost-effectiveness compared with human graders, Ophthalmology 124 (3) (March 2017) 343–351.

[20] K. Solanki, C. Ramachandra, S. Bhat, M. Bhaskaranand, M.G. Nittala, S.R. Sadda, EyeArt: automated, high-throughput, image analysis for diabetic retinopathy screening, Investigative Ophthalmology & Visual Science 56 (7) (June 2015) 1429.

[21] A.A. Jammal, A.C. Thompson, E.B. Mariottoni, S.I. Berchuck, C.N. Urata, T. Estrela, S.M. Wakil, V.P. Costa, F.A. Medeiros, Human versus machine: comparing a deep learning algorithm to human gradings for detecting glaucoma on fundus photographs, American Journal of Ophthalmology 211 (March 2020) 123–131.

[22] A.C. Thompson, A.A. Jammal, S.I. Berchuck, E.B. Mariottoni, F.A. Medeiros, Assessment of a segmentation-free deep learning algorithm for diagnosing glaucoma from optical coherence tomography scans, JAMA Ophthalmology 138 (4) (April 2020) 333–339.

[23] Q. Yan, D.E. Weeks, H. Xin, A. Swaroop, E.Y. Chew, H. Huang, Y. Ding, W. Chen, Deep-learning-based prediction of late age-related macular degeneration progression, Nature Machine Intelligence 2 (2) (February 2020) 141–150.

[24] U. Schmidt-Erfurth, S.M. Waldstein, S. Klimscha, A. Sadeghipour, X. Hu, B.S. Gerendas, A. Osborne, H. Bogunovic, Prediction of individual disease conversion in early AMD using artificial intelligence, Investigative Ophthalmology & Visual Science 59 (8) (July 2018) 3199–3208.

[25] R.T. Yanagihara, C.S. Lee, D.S. Ting, A.Y. Lee, Methodological challenges of deep learning in optical coherence tomography for retinal diseases: a review, Translational Vision Science & Technology 9 (2) (January 2020) 11.

5

Transfer learning-based detection of retina damage from optical coherence tomography images

Bam Bahadur Sinha[1], Alongbar Wary[2], R. Dhanalakshmi[3], K. Balakrishnan[3]

[1]DEPARTMENT OF COMPUTER SCIENCE AND ENGINEERING, INDIAN INSTITUTE OF INFORMATION TECHNOLOGY, RANCHI, JHARKHAND, INDIA; [2]SCHOOL OF COMPUTER SCIENCE AND ENGINEERING, VELLORE INSTITUTE OF TECHNOLOGY—AP UNIVERSITY, AMARAVATI, ANDHRA PRADESH, INDIA; [3]DEPARTMENT OF COMPUTER SCIENCE AND ENGINEERING, INDIAN INSTITUTE OF INFORMATION TECHNOLOGY, TIRUCHIRAPPALLI, TAMIL NADU, INDIA

5.1 Introduction

Diabetic retinopathy is a common cause of blindness due to damage to the retina's blood vessels [1]. Those who have had diabetes for more than 20 years have quite an 80% chance of developing the disease [2]. Diabetic retinopathy must be detected and diagnosed as early as possible. Diabetic retinopathy may cause up to 90% less blindness if properly treated and screened on a regular basis. Diagnostics for diabetic retinopathy are made by analyzing retinal scans and images. Ultrasonography, optical coherence tomography (OCT), and funduscopy (or ophthalmoscopy) are the most often used imaging techniques for this purpose. A fundus camera is used to take a picture of the backside of an eye, including macula, optic disc, and retina. For the diagnosis of diabetic retinopathy, ultrasonography employs ultrasound to develop an image. Due to its noninvasive imaging approach that provides micrometer-resolution cross-sections of the retina, OCT has revolutionized eye disease detection regardless of the capabilities of fundoscopic and ultrasound imaging techniques.

It is well acknowledged that retinal diseases are among the primary reasons of acute vision loss on a worldwide scale and that they are receiving extensive and widespread attention as a result of this recognition. When it comes to the monitoring and diagnosis of disorders related to the retina, OCT is regarded to be a critical component. OCT [3] is a noninvasive imaging procedure that employs waves of light to create cross-sectional pictures of the retina [2]. Conventionally, clinicians analyze every cross-section inside the OCT separately, which increases the burden of the Optometrists and further leads to a high rate of misdiagnosis, as well as a limit on the ability to produce a substantial percentage of ground truth. The development of automated diagnostic techniques for retinal illnesses is proving to be very beneficial, as it allows ophthalmologists to be more effective in the evaluation and treatment of eye problems in their patients. Such automated diagnosis may be accomplished via the use of classic machine learning techniques [4] or deep learning techniques [5], which can be further empowered by different optimization techniques [6].

In the past few years, OCT has emerged as a strong imaging technique for noninvasive evaluation of a wide range of retinal abnormalities, such as helping in the detection of DRUSEN, CNV, and DME, as seen in Fig. 5.1. However, as the quantity of image data generated by modern OCT equipment continues to grow, the viability of traditional manual OCT evaluation in clinical settings has become more unfeasible, if not completely impractical. Similarly, when accomplished by experienced ophthalmologists, the assessment of retinal diseases is concerned with significant intra- and interobserver variability, which can lead to erratic and unreliable analysis, slowing accurate diagnosis and placing a strain on available healthcare resources. The auto-diagnosis of retinal problems in OCT images would be very beneficial, as it would allow ophthalmologists to quickly examine and treat retinal disease.

Deep learning (DL) approaches have made significant progress in the state of the art for healthcare image identification, and this is expected to continue. Deep learning, which comes under the domain of artificial intelligence technology, mimics the functioning of a human brain in terms of processing the data and creating patterns that can be used in decision-making [7]. Deep learning is a technique that is becoming more popular. When DL performs a challenging diagnosis for optometrists professionals utilizing a large number of images, it marks a watershed moment in the illness diagnostic

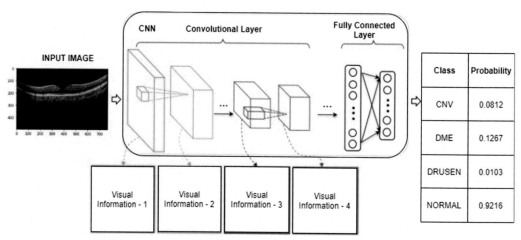

FIGURE 5.1 Working principle of CNN.

process. For training purposes, techniques such as convolutional neural networks (CNN) often need a large quantity of data. Occasionally, the amount of data available for the training process is insufficient. CNNs are thought to be hampered in their use in practical applications due to a lack of data. To address this issue, transfer learning (TL) was developed and implemented. The fundamental notion of TL is the transfer of information from one to another domain to enhance an operation. TL necessitates the process of fine-tuning the CNN model while making only minor modifications. The impracticality of manual assessment of diabetic retinopathy has been attributed to concerns such as restricted access to professionals and an increase in patient count. Development of an automated technique of diabetic retinopathy diagnosis to aid eye-care providers to circumvent these restrictions, for example [8]. In this case, deep learning is being employed, which has been shown to provide excellent results in a variety of image classification disciplines [9]. Deep learning along with transfer learning is being applied in this case that yields promising results. One of the benefits of deep learning is the capability of the network to learn itself and to map low-level to high-level image properties at various levels, which allows it to classify images more accurately and efficiently [10]. Conventional machine learning approaches, on the other hand, need the creation of image features by hand. Deep learning models, despite their many benefits, are time consuming to develop and need a considerable quantity of labeled data to be used in the training process.

Using TL, which involves the use of a previously trained model to perform tasks that are totally different from the initial job for which the model was taught, it is possible to avoid the aforementioned problem entirely. Transfer learning has been found to be quite successful, particularly in situations when vast amounts of data are unavailable and minimal computing costs are required.

In this chapter, use of transfer learning for the goal of identifying diabetic retinopathy using OCT images is used. First, we use our data to retrain many existing pretrained models, and then we choose the model that performs the best overall. Afterward, an investigation is performed on this model to better accomplish our goal, and we use the optimized network to retrieve image attributes. Following that, we employ these attributes to train a VGG16 convolutional network and an Inception V3 convolutional network, which are then used to complete the final classification job of distinguishing diabetic retinopathy. The remaining section of the chapter is structured as follows: Section 2 highlights the related background, and Section 3 discusses the different methodologies used in this chapter. Section 4 illustrates the flow of the proposed model followed by the obtained results in Section 5. The closing Section 6 concludes the chapter with future direction.

5.1.1 Related background

Machine learning and deep learning approaches have boosted the state of the art in terms of image recognition in the medical domain over the course of many years.

Ref. [11] proposed a method for detection of which was formulated on derived local binary sequence attributes from optical coherence tomography images and lexicon learning utilizing bag of words modeling techniques. The method had a sensitivity and specificity of 81.2% and 93.7%. Ref. [12] used an ensemble method to categorize retina (healthy and unhealthy) using retinal characteristics from SD-OCT images, and the results showed that the approach had a mean correctness of more than 96%. Ref. [13] developed a linear SVM to categorize normal retina and DME and found that it has 87.5% sensitivity and 87.5% specificity. Ref. [14] used HOG descriptors and SVM classification algorithms to detect AMD, DME, and normal retina, with an overall accuracy of around 100%, 100%, and 86.67%, respectively, at the OCT level, according to the results of his study. This was due to the strong reliance on characteristics expressly specified by ophthalmologists utilizing their domain expertise, which resulted in a time-consuming process with poor generalization ability and, in some cases, impossibility in huge datasets.

With the use of OCT images, the researchers examined the effects of transfer learning and a restricted framework on eye disease. In comparison to the restricted model, which obtained an overall of 93.4%, 96.6% true positive rate, 94.0% true negative rate, and 12.7% weighted error, it attained 96.6% accuracy, 97.8% true positive rate, 97.4% true negative rate, and 6.6% weighted error using transfer learning [15]. Ref. [16] proposed an additional transfer learning model based on AlexNet. After being trained using ImageNet to extract characteristics from OCT pictures, the proposed model is used to differentiate between the normal and DME eyes using the SVM classifier (binary classifier). It obtained 97.66% precision. Using and training DOCTAD, the researchers in Ref. [17] were able to categorize the individual optical coherence tomography images as atrophied or normal and forecasted the defects in defective regions in the volumes of OCT. There was an application of a TL approach for demonstrating the advantages of learning from a subject's previous search for the information to further improvise segmentation at a later stage. To produce a classifier model for retinal eye illnesses, Ref. [18] conducted a comparison research between training a CNN and utilizing a TL-based deep learning model. With the CNN, they obtained 100% true positive rate, 91.8% of true negative rate, and 99.0% accuracy, whereas with a TL-based deep learning model, they obtained a true positive rate of 98%, true negative rate of 87.3%, and 93.9% accuracy.

Several factors influence the effectiveness of TL approaches for diagnosis and prognosis of retinal disease. These factors include the pretrained DL model, training images count and quality, type, the steps taken during preprocessing, image attributes (e.g., color, size), the extent of difference between the actual and target training databases, and classes. Optimization of such parameters has the potential to increase performance. Also useful for this reason will be the new network designs that are being designed to handle a variety of activities. Accordingly, more research on transfer learning for diabetic retinopathy diagnostics is necessary and recommended. Table 5.1 demonstrates a few recent findings, approaches, and performance of different models proposed by several researchers in the direction of classifying diabetic retinopathy.

Table 5.1 Performance of existing models for classifying diabetic retinopathy.

Reference	Technique/model	Dataset	Model performance
[19]	Multi-ResNet50	Proprietary	Accuracy: 97.9%, AUC: 0.998
[20]	LACNN	UCSD	Accuracy: 90.01%, AUC: 0.9940
[21]	VGG-16	Proprietary	Accuracy: 98.6%, AUC: 1.0
[22]	DenseNet	Proprietary	Accuracy: 94.5%, AUC: 0.9921
[15]	Inception V3	Public	Accuracy: 96.6%, AUC:0.999
[12]	Random Forest	Public	Accuracy: 96%, AUC:0.99
[23]	MCME	NEH dataset	AUC: 0.998
[24]	Deep CNN—14 layer	Public (KMC)	Accuracy: 91.17%
[25]	Deep neural network	Proprietary	Accuracy: 95.9%, AUC: 0.984

5.2 Experimental methodology

In this section, we discuss different methodologies used in this chapter for detecting retina damage from optical coherence tomography images. We have used the deep learning model with transfer learning to do the same. Following subsection highlights the working of the VGG16 convolutional network and Inception V3 convolutional network.

5.2.1 VGG16 convolutional network

The VGG model, commonly known as VGGNet, that actually supports 16 layers is often alluded to as VGG16. It is a CNN model that was developed by Ref. [26]. VGGNets are built on CNN's most significant features (CNN). The fundamental principle underlying how a CNN works is illustrated via Fig. 5.1. Each layer in a CNN creates a new interpretation for an input image by progressively collecting discriminative information from the image over time. It can be noted that irrelevant information is progressively deleted from lower levels and relocated to upper layers as the hierarchy progresses. The reconstructions of the final layer only maintain the portions that are the most discriminatory.

Comparatively small convolutional filters are used to build the VGG network. In the design architecture of VGG-16, it comprises three fully connected layers and 13 convolutional layers. The convolutional layers of VGG use a modest receptive field (3×3), the smallest size that somehow still catches (up or down) and (left or right). In addition, there are 1×1 convolution filters that operate as a linear alteration of the input. Then, there is a ReLU unit, which is a significant AlexNet development that cuts training time in half. The ReLU is a linear function that gives an output as input in case the input is positive, otherwise, it yields an output of 0. To maintain spatial resolution after convolution, the convolution stride is set to 1 pixel. The VGG network's hidden layers leverage ReLU. Local response normalization (LRN) is seldom used in VGG as it increases

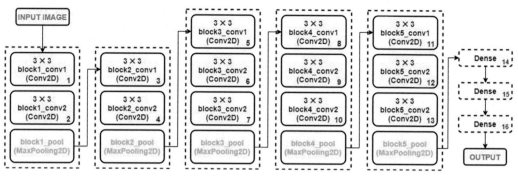

FIGURE 5.2 VGG16 architecture.

memory usage and training time. Furthermore, it has no effect on total accuracy. Fig. 5.2 illustrates the architecture of the VGG16 convolutional network.

The number 16 in the term VGG alludes to the deep neural network (DNN) 16 layers (VGGnet). This indicates that VGG16 is a large network with over 138 million parameters. Even by contemporary standards, it is a massive network. The simplicity of the VGGNet16 design, on the other hand, ultimately renders the network more enticing. It may be argued that its architecture is fairly uniform just by glancing at it. A few convolution layers are preceded by a pooling layer that diminishes the height and breadth of the image. When it comes to the amount of filters we can employ, we have roughly 64 filters, which we can expand to around 128 and subsequently to 256. We may utilize 512 filters in the final layers. A number of filters that we may apply increases by a factor of two with each step or with each layer of the convolution layer. This is a fundamental notion that guided the creation of the VGG16 network's architecture. It is important to note that the VGG16 network has some significant drawbacks, one of which is that it is a large network, which means that it requires a longer time for training its parameters.

5.2.2   Inception V3 convolutional network

The Inception v3 model, which was launched in 2015, features 42 layers and a reduced error rate compared to its predecessors. The following are the various improvements that strengthen the inception V3 model:

- Factorization of larger convolutions into smaller ones.
- Factorization of spatial data into asymmetric convolutions.
- To handle the vanishing gradient problem present in DNN, utilization of auxiliary classifiers is promoted.
- Reduction in the size of the grid.

The Inception V1 architecture has a significant advantage in terms of size reduction. The model was further improved by factorizing the larger convolutions into smaller convolutions. Consider the inception V1 basic structure. Suppose it contains a 5 × 5 convolutional layer, which, as previously said, was relatively expensive to compute. As a result, the 5 × 5 convolutional layer was transformed into two 3 × 3 convolutional layers to lower the computational cost. By using two 3 × 3 convolutions, the parameters are lessened. The lower number of parameters also results in a decrease in computing expenses. Despite the fact that the larger convolutions are being factored into smaller ones. You may be wondering what would happen if we could factorize further, say to a 2 × 2 convolution. However, asymmetric convolutions were a more effective way to increase the model's efficiency. Convolutions that are asymmetric in nature have the form $n \times 1$. Consequently, they substituted a (1 × 3) convolution for the (3 × 3) convolutions, accompanied by a (3 × 1) convolution. This is equivalent to shifting a 2-layer network having the identical receptive field as a 3 × 3 convolutional network. The purpose of utilizing an auxiliary classifier would be to boost the convergence rate of extremely DNNs. The auxiliary classifier is primarily used to address the issue of vanishing gradients in extremely deep networks. The auxiliary classifiers had little effect on performance during the early phases of training. However, in the ending, the model with auxiliary classifiers outperformed the model without auxiliary classifiers in terms of accuracy. Therefore, in the Inception V3 model, the auxiliary classifiers work as a regularizer. Formerly, max and average pooling were employed to minimize the grid size of feature space. To effectively minimize the grid size in the inception V3, the activation dimension is extended. Fig. 5.3 [27] illustrates the architecture of Inception V3 model.

5.2.3   Xception model

The DNN design that utilizes depthwise separable convolutions (DSC) is referred as Xception model. Researchers at Google developed it. Inception modules in CNNs have

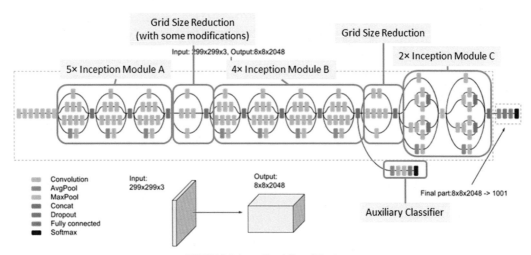

FIGURE 5.3 Inception V3 architecture.

been interpreted by Google as an intermediary stage between ordinary convolution and DSC operation. DSC might be thought of as inception modules with infinitely numerous towers. As a result of this discovery, they propose a revolutionary deep CNN architecture based on a modified version such that the inception modules got replaced with DSCs.

Fig. 5.4 [28] illustrates the Xception network. The data initially pass via the entering flow, then eight times through the middle flow, and lastly through the exit flow. The batch normalization is performed on all separable convolution and convolutional layers. To be efficient, XCeption's architecture depends on the following two factors:

- Classical convolutions may be replaced with DSCs, which are said to be substantially more efficient in terms of computing time than the formers.
- Similar to what can be observed in ResNets, Xception models maintain shortcuts within convolutional blocks.

DSCs are further divided into two steps: depthwise and pointwise convolution. Depthwise convolution is an initial step in which, rather than using a $(d \times d \times C)$ convolution size, we use a $(d \times d \times 1)$ convolution. We do not perform convolution on all channels simultaneously, but rather one by one. Pointwise convolution performs a standard convolution with size $(1 \times 1 \times N)$ across the volume $(K \times K \times C)$. As stated

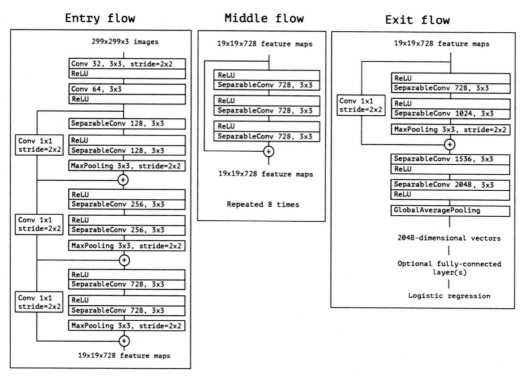

FIGURE 5.4 Xception architecture.

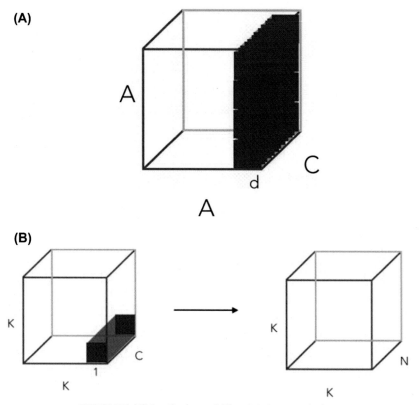

FIGURE 5.5 (A) Depthwise and (B) pointwise convolution.

before, this enables the creation of a volume of form $(K \times K \times N)$. Fig. 5.5 gives an illustration of depthwise and pointwise convolution process.

5.3 Proposed model

This section of the chapter discusses the flow diagram of the framework proposed for detecting retinal disease. The flow diagram can be broken down into four stages. The first stage deals with exploring the OCT images dataset. This stage gives insights regarding the different categories of OCT images and other details regarding the same. The second stage splits the dataset into a training and test set, such that the training set is used for training the model. The training set is further bifurcated into (training: validation). The test set is being used at the end to test the efficiency of the proposed model. The third stage aims at training the model using transfer learning and deep convolution network for classifying the OCT images appropriately. The final stage measures the efficacy of the proposed model using difference performance metrics such as accuracy, recall, precision, and F1-score. The four stages of our proposed model are illustrated in Fig. 5.6. The OCT image dataset [29], which has been used for detecting

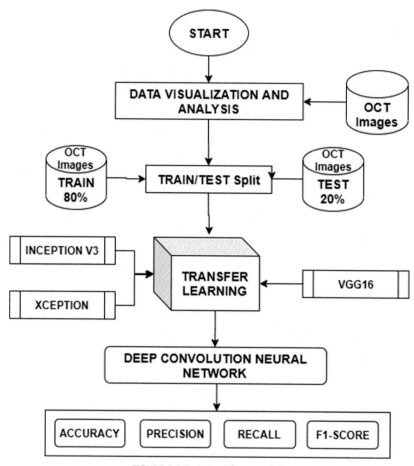

FIGURE 5.6 Proposed flow model.

retinal disease and for testing the proposed model performance, is publicly available on Kaggle. The results obtained at the output of each stage are further discussed in the upcoming section.

5.4 Experimental results and observations

The OCT image dataset used in our chapter comprises OCT images that belong to four different categories namely: DRUSEN, DME, CNV, and NORMAL. The experiment performed in Fig. 5.7 shows a sample image belonging to each category. Further after exploring the min/max pixel values of the images present in the dataset, it is observed

FIGURE 5.7 Categories of OCT images.

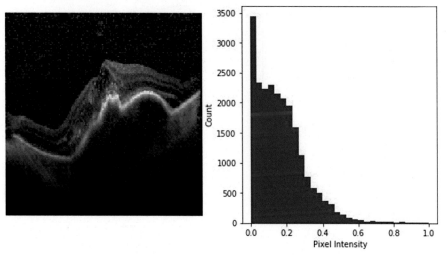

FIGURE 5.8 Histogram of RGB pixel intensities.

that the pixel values are already scaled between 0 and 1. This helps in efficient modeling with less complexity. Fig. 5.8 illustrates the histogram of RGB pixel intensities. During the implementation of the proposed model, the different categories of images are labeled as follows: Normal: 0; CNV: 1; DME: 2; DRUSEN: 3.

There is a clear sign of data imbalance that can be observed from Fig. 5.9 (left). To avoid that we have used performed undersampling. After doing the undersampling of the images, the issue of class imbalance is resolved as can be observed from Fig. 5.9 (right). Fig. 5.9 reflects the imbalance in the sample dataset of OCT images. Same can be visualized for complete training set as well. Oversampling along with data augmentation is a preferable solution as compared to undersampling but the undersampling process is faster as compared to oversampling. In future, we can make use of SMOTE approach to get rid of the issue of class imbalance. Our training dataset is quite large and suffers from class imbalance problem. It is utmost necessary to either perform undersampling or oversampling for balancing the classes. We have divided our OCT dataset into 80/20 ratio, where 80% is the training data, 20% is validation data. Table 5.2 gives the description about the number of images available

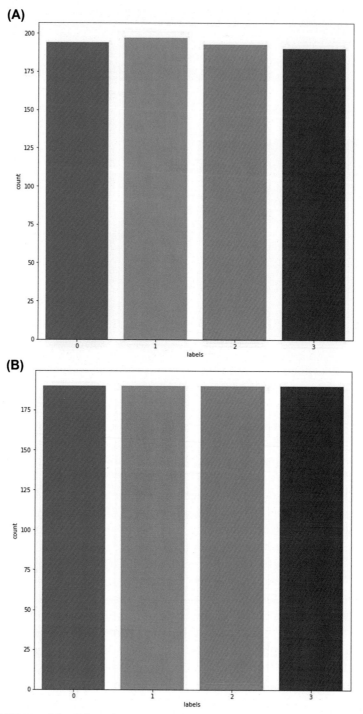

FIGURE 5.9 (A) Imbalanced test dataset and (B) test dataset after undersampling.

Table 5.2   Train-test split of OCT image dataset.

Category	Training set	Test set
CNV	37,208	242
DME	11,308	242
DRUSEN	8624	242
NORMAL	26,308	242

in training and test set. The next stage of the proposed model deals with diagnosing the retinal disease using transfer learning and convolutional deep neural network. Fig. 5.10 shows the obtained model accuracy and model loss over different number of epochs using transfer learning with VGG16 convolutional network. The diagnostic

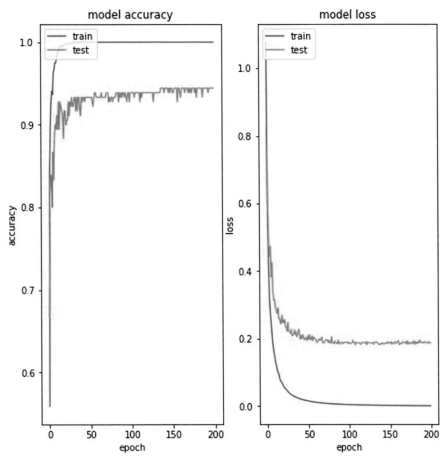

FIGURE 5.10 Model performance—transfer learning with VGG16 convolutional network.

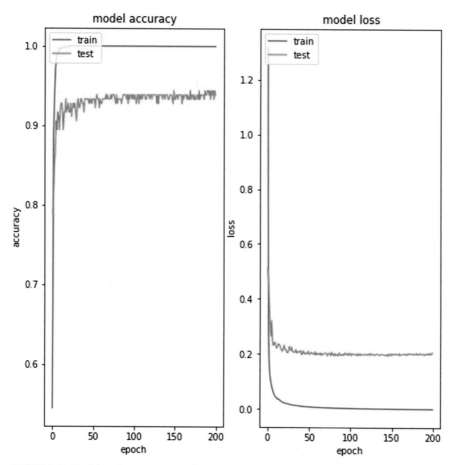

FIGURE 5.11 Model performance—transfer learning with Inception V3 convolutional network.

model is also tested using transfer learning with Inception V3 convolutional network. The accuracy and loss obtained by Inception V3 based model is illustrated via Fig. 5.11. The behavior of Xception model showed a nonuniform graph while testing on the test dataset. The model accuracy and loss obtained by transfer learning with Xception model is illustrated via Fig. 5.12.

The comparison of model performance in terms of accuracy, recall, precision and F1-score is demonstrated via Table 5.3. The confusion matrix obtained by each of the aforementioned models are also discussed. It can be clearly observed that transfer learning with VGG16 convolutional network and Xception model yields best performance. It even converges faster as compared to Inception V3 model.

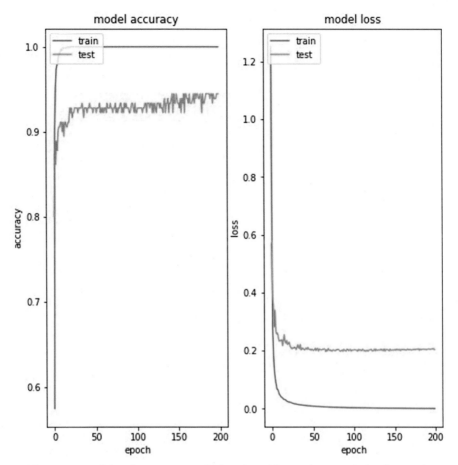

FIGURE 5.12 Model performance—transfer learning with Xception convolutional network.

5.5 Conclusion

In this chapter, a model based on TL is built on top of deep networks such as the Inception V3, VGG16, and the Xception network. The proposed framework was employed for the purpose of categorizing three retinal disorders and one normal class utilizing OCT images, and the results were promising. While compared to current state-of-the-art models for the diagnosis of retinal eye disorders, the suggested model attained the highest accuracy after leveraging a training dataset during the training process. This study shows that even a modest dataset may be used to achieve high performance for the multiclass categorization of different retinal diseases, which was previously thought to be impossible. Transfer learning in conjunction with deep convolutional networks was

Table 5.3 Performance comparison of transfer learning based models.

Model name	Performance

VGG16 Convolutional Network

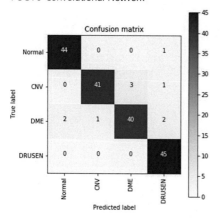

	precision	recall	f1-score	support
Normal	0.96	0.98	0.97	45
CNV	0.98	0.91	0.94	45
DME	0.93	0.89	0.91	45
DRUSEN	0.92	1.00	0.96	45
avg / total	0.95	0.94	0.94	180

Inception V3 Convolutional Network

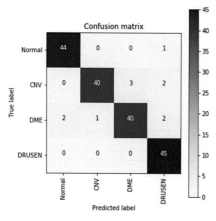

	precision	recall	f1-score	support
Normal	0.96	0.98	0.97	45
CNV	0.98	0.89	0.93	45
DME	0.93	0.89	0.91	45
DRUSEN	0.90	1.00	0.95	45
avg / total	0.94	0.94	0.94	180

Xception Convolutional Network

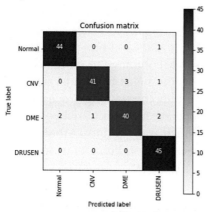

	precision	recall	f1-score	support
Normal	0.96	0.98	0.97	45
CNV	0.98	0.91	0.94	45
DME	0.93	0.89	0.91	45
DRUSEN	0.92	1.00	0.96	45
avg / total	0.95	0.94	0.94	180

used to solve the multiclassification problem in this chapter, and the results revealed the usefulness and advantages of doing so. Our approach has the potential to play a significant role in robustly nudging ailments and patient monitoring in a field that is heavily influenced by imaging, such as opthalmology, and to aid in the development of large-scale detection methods for eye diseases through quick, unbiased, and reliable image analysis.

When it comes to object recognition in the foreseeable future, semantic segmentation may be utilized to concentrate on one specific object instead of using thousands of tiny pixels to represent objects. As far as biological research are concerned, semantic segmentation is a relatively new motivator that has the potential to be very beneficial.

References

[1] W. Wang, A.C. Lo, Diabetic retinopathy: pathophysiology and treatments, International Journal of Molecular Sciences 19 (6) (2018) 1816.

[2] J. Tang, T.S. Kern, Inflammation in diabetic retinopathy, Progress in Retinal and Eye Research 30 (5) (2011) 343−358.

[3] D. Huang, E.A. Swanson, C.P. Lin, J.S. Schuman, W.G. Stinson, W. Chang, M.R. Hee, T. Flotte, K. Gregory, C.A. Puliafito, et al., Optical coherence tomography, Science 254 (5035) (1991) 1178−1181.

[4] B. B. Sinha, R. Dhanalakshmi, Building an adaptive recommendation model based on fuzzy mp neuron and weighted similarity indicator., Journal of Information Science and Engineering 37 (2).

[5] R.T. Yanagihara, C.S. Lee, D.S.W. Ting, A.Y. Lee, Methodological challenges of deep learning in optical coherence tomography for retinal diseases: a review, Translational Vision Science and Technology 9 (2) (2020), 11−11.

[6] B.B. Sinha, R. Dhanalakshmi, Evolution of recommender paradigm optimization over time, Journal of King Saud University-Computer and Information Sciences 34 (4) (2019) 1047−1059.

[7] B.B. Sinha, R. Dhanalakshmi, Building a fuzzy logic-based artificial neural network to uplift recommendation accuracy, The Computer Journal 63 (11) (2020) 1624−1632.

[8] S. Kuwayama, Y. Ayatsuka, D. Yanagisono, T. Uta, H. Usui, A. Kato, N. Takase, Y. Ogura, T. Yasukawa, Automated detection of macular diseases by optical coherence tomography and artificial intelligence machine learning of optical coherence tomography images, Journal of Ophthalmology 2019 (2019), 6319581, https://doi.org/10.1155/2019/6319581.

[9] J. Lee, Y.K. Kim, K.H. Park, J.W. Jeoung, Diagnosing glaucoma with spectral-domain optical coherence tomography using deep learning classifier, Journal of Glaucoma 29 (4) (2020) 287−294.

[10] B.B. Sinha, R. Dhanalakshmi, Building a fuzzy logic-based mcculloch-pitts neuron recommendation model to uplift accuracy, The Journal of Supercomputing 77 (3) (2021) 2251−2267.

[11] G. Lemaitre, M. Rastgoo, J. Massich, C.Y. Cheung, T.Y. Wong, E. Lam- oureux, D. Milea, F. Meriaudeau, D. Sidibe, Classification of sd-oct volumes using local binary patterns: experimental validation for dme detection, Journal of Ophthalmology 2016 (2016), 3298606. https://doi.org/10.1155/2016/3298606.

[12] M.A. Hussain, A. Bhuiyan, C.D. Luu, R. Theodore Smith, R.H. Guymer, H. Ishikawa, J.S. Schuman, K. Ramamohanarao, Classification of healthy and diseased retina using sd-oct imaging and random forest algorithm, PLoS One 13 (6) (2018) e0198281.

[13] K. Alsaih, G. Lemaitre, M. Rastgoo, J. Massich, D. Sidibe, F. Meriaudeau, Machine learning techniques for diabetic macular edema (dme) classification on sd-oct images, BioMedical Engineering Online 16 (1) (2017) 1−12.

[14] P.P. Srinivasan, L.A. Kim, P.S. Mettu, S.W. Cousins, G.M. Comer, J.A. Izatt, S. Farsiu, Fully automated detection of diabetic macular edema and dry age-related macular degeneration from optical coherence tomography images, Biomedical Optics Express 5 (10) (2014) 3568–3577.

[15] D.S. Kermany, M. Goldbaum, W. Cai, C.C. Valentim, H. Liang, S.L. Baxter, A. McKeown, G. Yang, X. Wu, F. Yan, et al., Identifying medical diagnoses and treatable diseases by image-based deep learning, Cell 172 (5) (2018) 1122–1131.

[16] G.C. Chan, A. Muhammad, S.A. Shah, T.B. Tang, C.-K. Lu, F. Meri- audeau, Transfer learning for diabetic macular edema (dme) detection on optical coherence tomography (oct) images, in: 2017 IEEE International Conference on Signal and Image Processing Applications (ICSIPA), IEEE, 2017, pp. 493–496.

[17] J. Loo, L. Fang, D. Cunefare, G.J. Jaffe, S. Farsiu, Deep longitudinal transfer learning-based auto- matic segmentation of photoreceptor ellipsoid zone defects on optical coherence tomography images of macular telangiectasia type 2, Biomedical Optics Express 9 (6) (2018) 2681–2698.

[18] Arrigo, F. Romano, E. Aragona, C. Di Nunzio, M. Battista, F. Bandello, M.B. Parodi, Optical coherence tomography angiography can categorize different subgroups of choroidal neo- vascularization secondary to age-related macular degeneration, Retina 40 (12) (2020) 2263–2269.

[19] F. Li, H. Chen, Z. Liu, X.-d. Zhang, M.-s. Jiang, Z.-z. Wu, K.-q. Zhou, Deep learning-based automated detection of retinal diseases using optical coherence tomography images, Biomedical Optics Express 10 (12) (2019) 6204–6226.

[20] L. Fang, C. Wang, S. Li, H. Rabbani, X. Chen, Z. Liu, Attention to lesion: lesion-aware convolutional neural network for retinal optical coherence tomography image classification, IEEE Transactions on Medical Imaging 38 (8) (2019) 1959–1970.

[21] F. Li, H. Chen, Z. Liu, X. Zhang, Z. Wu, Fully automated detection of retinal disorders by image-based deep learning, Graefes Archive for Clinical and Experimental Ophthalmology 257 (3) (2019) 495–505.

[22] J. De Fauw, J.R. Ledsam, B. Romera-Paredes, S. Nikolov, N. Tomasev, S. Blackwell, H. Askham, X. Glorot, B. O'Donoghue, D. Visentin, et al., Clinically applicable deep learning for diagnosis and referral in retinal disease, Nature Medicine 24 (9) (2018) 1342–1350.

[23] R. Rasti, H. Rabbani, A. Mehridehnavi, F. Hajizadeh, Macular oct classification using a multi-scale convolutional neural network ensemble, IEEE Transactions on Medical Imaging 37 (4) (2017) 1024–1034.

[24] J.H. Tan, S.V. Bhandary, S. Sivaprasad, Y. Hagiwara, A. Bagchi, U. Raghavendra, A.K. Rao, B. Raju, N. S. Shetty, A. Gertych, et al., Age-related macular degeneration detection using deep convolutional neural network, Future Generation Computer Systems 87 (2018) 127–135.

[25] W. Lu, Y. Tong, Y. Yu, Y. Xing, C. Chen, Y. Shen, Deep learning-based automated classification of multi-categorical abnormalities from optical coherence tomography images, Translational Vision Science and Technology 7 (6) (2018), 41-41.

[26] K. Simonyan, A. Zisserman, Very Deep Convolutional Networks for Large- Scale Image Recognition, arXiv preprint arXiv:1409.1556.

[27] C. Szegedy, V. Vanhoucke, S. Ioffe, J. Shlens, Z. Wojna, Rethinking the inception architecture for computer vision, in: Proceedings of the IEEE Conference on Computer Vision and Pattern Recognition, 2016, p. 28182826.

[28] F. Chollet, Xception: deep learning with depthwise separable convolutions, in: Proceedings of the IEEE Conference on Computer Vision and Pattern Recognition, 2017, pp. 1251–1258.

[29] Kermany, K. Zhang, M. Goldbaum, et al., Labeled optical coherence tomography (oct) and chest x- ray images for classification, Mendeley Data 2(2).

6

An improved approach for classification of glaucoma stages from color fundus images using Efficientnet-b0 convolutional neural network and recurrent neural network

Poonguzhali Elangovan, D. Vijayalakshmi, Malaya Kumar Nath

DEPARTMENT OF ECE, NATIONAL INSTITUTE OF TECHNOLOGY PUDUCHERRY, THIRUVETTAKUDY, KARAIKAL, PUDUCHERRY, INDIA

6.1 Introduction

The human visual system is responsible for visualizing and interpreting the objects around us. Human eye is the most complex sense organ, which gathers light from the surrounding environment and converts it to nerve impulses. These vision-related sensory signals are transmitted to the brain by optic nerve. The brain generates an image and thus provides vision. The amount of light entering the eye through the pupil, shape of the pupil, shape of the eyeball, and the appearance of the lens can all have an impact on eyesight [1]. Thus, for a healthy eye, good eyesight and better vision are required. In general, vision loss is any reduction in the ability to perceive objects. It can affect one or both eyes and may occur suddenly or develop gradually over time, may be partial or complete. Any inflammation to the specific parts of the eye such as the lens, retina, macula, and optic nerve may result in curable or noncurable vision problems. Some common ocular diseases that may affect human vision are glaucoma, age-related macular degeneration, diabetic retinopathy, cataract, and retinitis pigmentosa. According to the world health organization vision study [2], it is estimated that the number of visually impaired people worldwide is 2.2 billion, with at least 1 billion cases could have been prevented or are still unaddressed. Hence, early identification and continuous monitoring are essential to prevent vision loss.

Optic nerve comprises more than one million nerve fibers and is situated at the back of the eye. They carry visual information from the eyes to the brain. The deterioration of the optic nerve may result in various ocular disorders like glaucoma, optic nerve drusen, optic nerve atrophy, and optic neuritis. Glaucoma, also referred to as sneaky robber of vision, is the second main cause of blindness worldwide. It is a cluster of ocular conditions caused due to elevated intraocular pressure. The major cause of elevated intraocular pressure (IOP) is increased aqueous fluid circulation at the pupil, which is caused by a greater barrier to the aqueous fluid through the trabecular meshwork [1]. Glaucoma degrades the tunnel vision at the initial stages and the further progression will result in permanent blindness [3]. Fig. 6.1 depicts the sample photo viewed by a normal patient and glaucoma-infected patient. Due to the substantial loss of optic nerve fiber,

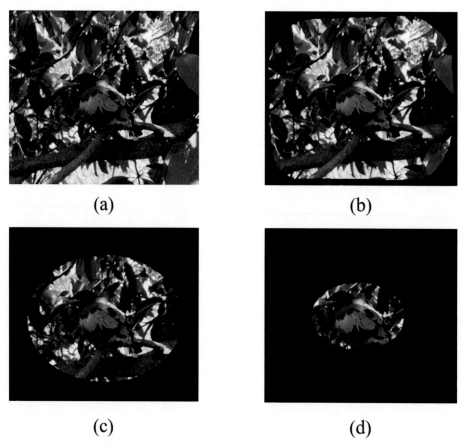

FIGURE 6.1 Visualization of the image in normal and glaucoma infected eye: (A) normal eye, (B) mild glaucoma affected eye, (C) advanced glaucoma affected eye, (D) extreme glaucoma affected eye.

peripheral vision is greatly affected. This is clearly depicted in various stages of glaucoma (such as mild, advanced, and extreme).

Primary, secondary, and congenital are the three main types of glaucoma. Depending on the anatomy of the anterior chamber angle, primary and secondary glaucoma are further classified into open-angle and angle-closure glaucoma. In primary open-angle glaucoma (POAG), the drainage angle of the eye is open, but it does not allow aqueous humor fluid to drain adequately. As a result, the eye's drainage system becomes clogged over time, causing a gradual increase in IOP. This type of glaucoma is painless and often goes unnoticed until the optic nerve is severely damaged. When the drainage angle of the eye becomes suddenly and completely blocked, IOP rises very quickly. This condition is referred to as a primary angle-closure glaucoma (PACG). Secondary glaucoma is caused by external factors like high incidence of ocular infections, inflammations, and complicated cataract surgery. Congenital glaucoma is a group of disorders characterized by abnormally high IOP caused by developmental abnormalities in the anterior chamber angle, which obstructs aqueous humor drainage. As the damage to the optic nerve caused by glaucoma is irreversible, it is therefore imperative to detect glaucoma early so that visual morbidity can be avoided. POAG, the most common type of glaucoma, develops slowly and is painless. If not treated, peripheral vision may deteriorate first, followed by central vision, eventually leading to blindness. Moreover, vision loss due to POAG, once it has occurred, is permanent.

Glaucoma screening is very ponderous because symptoms appear only in an advanced stage. This necessitates the precocious diagnosis of glaucoma. Clinical examinations of glaucoma need a skilled expert and may be susceptible to intra/interobserver errors, as it involves various control parameters. Various imaging modalities have been developed to assist the clinical diagnosis. However, these instruments are expensive and bulky. Furthermore, manually inspecting the individual images is time consuming. The fundus image obtained using a fundus camera is cost-effective and can aid in early detection in situations where more expensive equipment is unavailable. Fig. 6.2 depicts the fundus image of the healthy eye and glaucoma-affected eye. The region depicted under the red square box is referred as region of interest (ROI) for glaucoma detection. It comprises optic disc (OD), optic cup (OC), and vascular network. The optic disc appears as a bright circular or elliptical region and inner brighter region is optic cup. The cup size is small in the healthy disc, whereas an elongated optic cup is noticed in the glaucomatous disc [4].

Despite the fact that the pathological features of the optic disc are clearly visible in the fundus images, manual examination is a time-consuming process. With the rising number of persons affected by glaucoma, relying on these interpretations to diagnose the disease may become untenable. Hence, developing a computer-aided detection system for glaucoma diagnosis could yield effective results [5]. By automating the detection process for ocular diseases, CAD offers early detection and easy access for patients, thereby reducing the burden on clinicians. In general, CAD-based approach is grouped into two major types: machine learning (ML) based and deep learning (DL)

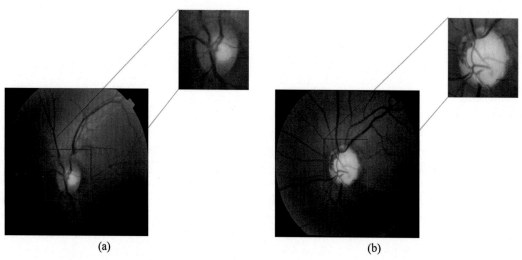

FIGURE 6.2 Fundus image highlighted with region of interest: (A) healthy, (B) glaucoma.

based. The performance of the ML-based approaches mainly relies on the hand-crafted features, which are obtained by applying various image processing techniques. DL-based approaches perform feature extraction and classification using a single model. This work aims to enhance the performance of glaucoma classification tasks using deep learning techniques.

The major objectives of this work are stated as follows:

- To propose an unique approach based on convolutional neural network and bidirectional long-short term memory (Bi-LSTM) network for the classification of glaucoma stages from fundus images.
- To investigate the effectiveness of various classifiers like softmax, SVM, KNN, and Bi-LSTM for classifying the stages of glaucoma.
- To explore the feature selection techniques (such as wavelet transform and principal component analysis) for glaucoma classification task.

The remainder of paper is structured as follows. Section 6.2 explains about the related work briefly. The database and suggested approach are described in Section 6.3. Findings and conclusions are discussed in Section 6.4 and Section 6.5, respectively.

6.2 Related work

The deep learning model learns hierarchical features from raw data and classifies them using a cascade of layers [6]. The remarkable success of convolutional neural networks (CNNs) has made them a promising choice for image classification tasks. Several algorithms have been developed in recent years for glaucoma detection using CNN. Authors have either developed a CNN from scratch [7–10], or employed the existing models. However, a brief summary of existing glaucoma detection schemes using transfer learning is discussed, as it is most useful for this research work. In some works [11–13], the model is trained with the region of interest (optic disc) of the image, as it provides the discriminative features crucial for classification. The rest of the work, on the other hand, takes into account the complete fundus image [14–21].

Fu et al. [11] have developed a disc-aware ensemble network based on an ensemble of four independent networks. The predictions obtained by the networks are fused to achieve the final decision. With a total of 1676 (46 glaucoma and 1630 normal) and 5783 (113 glaucoma and 5670 normal) images, authors have reported an area under the curve (AUC) of 0.918 and 0.817 in Singapore Chinese Eye Study (SCES) and Singapore Indian Eye Study (SINDI) databases, respectively. Christopher et al. [12] have demonstrated the impact of native and transfer learning of pretrained models for glaucoma detection. By analyzing 14,822 images, the authors have concluded that fine-tuning the Resnet-50 model outperformed the other two models, VGG-16 and Inception-v3, with an overall AUC of 0.91.

Investigation of VGG-19, Googlenet, and Resnet-50 is carried out by Ref. [15]. Around 2313 fundus images obtained from RIM-ONE, DRISHTI-GS1, and ESPERANZA are used to evaluate the models. With the highest AUC of 0.94, VGG19 network surpasses the other models. Ko et al. [19] have presented an ensemble approach based on VGG-16 and SVM classifier. First, the features extracted from VGG-16 using transfer learning are utilized to compute the confidence score. If the computed score is lower than a threshold value, SVM classifier is used. A total of 944 fundus images collected from the local hospital are used. With 187 test images, they have reported an accuracy of 92.8% in the ensemble model.

Performance comparison of Inception-v3, Resnet-50, Xception, VGG-16, and VGG-19 is discussed in Ref. [16]. Authors have experimented with 1707 images and reported that Xception model leads the other models with an average accuracy of 0.9605. Zhen et al. [17] have examined the behavior of Densenet, Inception-v3, InceptionResnet, NasnetMobile, Resnet, VGG-16, VGG-19, and Xception for glaucoma detection. Models are trained and tested with 5978 fundus images collected from the hospital. Densenet model has achieved the highest accuracy of 75.50%.

Liu et al. [13] have investigated the performance of Resnet-50 for glaucoma classification. Optic disc localization is performed at the preprocessing stage.

The efficacy of the model is examined using 3768 images acquired from three ophthalmic centers and 626 images from RIM-ONE and HRF database. Sensitivity of 89.3% and specificity of 97.1% are reported in their work. Jin et al. [20] have used a deep learning approach for the detection of early and advanced glaucoma. The pretrained Inception-v3 model is fine-tuned using a transfer learning approach. Training is done with 754 images (228 advanced, 141 early, and 385 normal). Authors have reported an accuracy of 84.5% with 464 images (141 advanced, 87 early, and 236 normal).

Pretrained Densetnet-201 model is employed in Ref. [18]. The fundus images from DRISHTI-GS1 database are preprocessed for better contrast enhancement. With 80% of images, the model is trained using Adam optimizer. An overall classification accuracy of 96.48% is obtained with 20% of test images. Rehman et al. [21] have presented an ensemble framework for glaucoma detection. Predictions obtained by four pretrained models such as Alexnet, Incepiton-v3, InceptionResnet-v2, and Nasnet-Large are combined using different voting techniques. Authors have concluded that ensembling the models using an accuracy-based weighted voting scheme outperformed with an overall classification accuracy of 99.6%, 88.2%, 95.1%, 96.1%, and 99.8% in ACRIMA, ORIGA, RIM-ONE, Armed Forces Institute of Opthalmology (AFIO), and Hayatabad Medical Complex (HMC) databases, respectively. An ensemble approach for the classification of glaucoma stages is presented in Cho et al. [14]. Modified versions of Inception-v3 and InceptionResnet-v2 are utilized in various strategies. Authors have obtained an improvement in the accuracy of about 3% in the ensemble method compared with the single best CNN model with 3460 fundus images.

Table 6.1 summarizes the existing deep learning-based CAD approach for glaucoma classification. The performance of pretrained models for the classification of normal and glaucoma pictures has been the focus of the majority of previous research. Moreover, private databases are utilized in most of the existing works. It is observed that only a limited work has been done on employing deep neural networks to classify glaucoma stages. This motivated us to develop an unique approach for the classification of normal, early glaucoma, and advanced glaucoma samples. Furthermore, numerous classifiers in various configurations are investigated to improve the overall performance of the suggested approach.

6.3 Methodology

Classification of early and advanced glaucoma stages from the fundus images may aid in providing the suitable medical treatment to the affected patients. In this work, the effectiveness of CNN and Bi-LSTM are integrated to categorize the fundus images into normal, early glaucoma, and advanced glaucoma.

6.3.1 Database

The database considered in this work comprises of 1544 (788 normal, 287 early glaucoma, and 467 advanced glaucoma) fundus images, obtained from Kims Eye Hospital [20]. All the images are acquired using a nonmydriatic auto fundus camera. The distribution of images in the database is depicted in Fig. 6.3. This dataset includes 51% of normal images, 19% of early glaucoma images, and 30% of advanced glaucoma images. The images are OD centered with a dimension of 240 × 240 in portable network graphic (PNG) format. The sample normal images and glaucoma images are depicted in Fig. 6.4. The elongated optic cup is noticed in early and advanced glaucoma stages.

Table 6.1   Various existing methods for glaucoma detection using CNN.

Author	Network architecture	Data augmentation	Preprocessing techniques	Optimizer	Database (Name, size)	Performance measures
Zhen (2017) [17]	Pretrained models	Rotation, flipping	Circular ROI detection	SGDM	5978 images from hospital	Highest ACC: 0.755 for densenet
Fu (2018) [11]	Disc-aware ensemble network	Rotation, cropping, flipping	Scaling	SGD	SCES—1676 SINDI—5783	AUC: 0.918 AUC: 0.817
Christopher (2018) [12]	Pretrained models	Horizontal mirroring, cropping, translation	Scaling	SGDM	14,822 images from hospitals	Highest AUC: 0.91 for Resnet-50
Liu (2018) [13]	Resnet-50	—	ROI extraction, scaling	SGDM	4000 images from hospital and RIM-ONE, HRF	ACC: 0.927 SEN: 0.893 SPE: 0.971
Juan (2019) [15]	Own CNN pretrained models	Flipping, rotation, translation	Optic disc localization ROI extraction scaling,	SGD	2313 images from DRISHTI-GS1 RIM-ONE ESPERANZA	Own CNN: AUC: 0.8969 SEE: 0.8703 BACC: 0.8303
Ko (2020) [19]	Ensemble model	Rotation, shifting, sheering, zooming	Gaussian blur ROI detection, Scaling	SGDM	944 from hospital, DRISHTI-GS1-101	ACC: 0.928 ACC: 0.803
Andres (2019) [16]	Pretrained models	Not applied	ROI extraction, scaling	SGD	1707 images from public databases	Highest ACC: 0.96 for xception
Rehman (2021) [21]	Ensemble model	Flipping, rotation, translation	ROI detection, Scaling	SGDM	ACRIMA ORIGA RIM-ONE AFIO HMC	ACC:0.996 0.882 0.951 0.961 0.998
Cho (2021) [14]	Ensemble model	Rotation, cropping	Artifact removal, Scaling	SGDM	3460 images from hospitals	Highest ACC: 0.941 for ensemble model

6.3.2   Proposed framework

The general pipeline of the proposed methodology is given in Fig. 6.5. It includes four major blocks: preprocessing, feature extraction, feature selection, and classification. Oversampling data-level techniques and image resizing are performed in the preprocessing stage. The feature extraction block extracts the relevant features from the images. This is followed by feature selection, where the prominent features are selected from the extracted features. Finally, the classification of various stages is performed in the classification block.

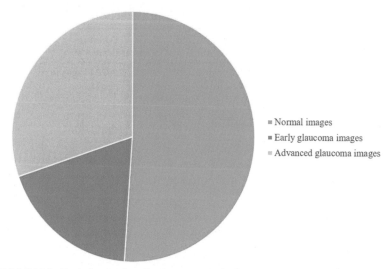

FIGURE 6.3 Distribution of normal, early glaucoma, and advanced glaucoma images in the dataset.

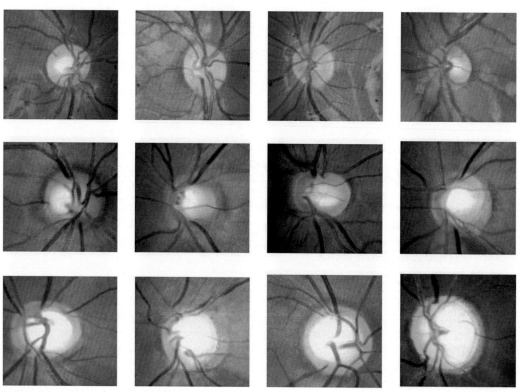

FIGURE 6.4 Sample normal (represented in the first row), early glaucoma (represented in the second row), and advanced glaucoma (represented in the third row) images.

FIGURE 6.5 Pipeline of the proposed approach for the classification of glaucoma stages from fundus images.

Preprocessing: Deep learning model, in general, performs well with large annotated data. The data distribution of the original database reveals the existence of class or data imbalance problem. Handling such problems is essential in training deep learning architectures. In this work, oversampling data-level strategy is employed to handle the data imbalance problem. To generate a balanced dataset, images are enlarged in all the classes using the rotation augmentation technique. Each image in the advanced glaucoma class is rotated by 0.5 steps between zero and 17. This results in a total of 16,345 images. As the early glaucoma class comprises less number of images (289), each image is rotated between zero and 11 with a step size of 0.2, yielding 16,184 images. A step size of 0.1 is employed while rotating each image in normal class between zero and 2. This generates a total of 16,584 images.

Feature extraction: In this work, the pertinent features are extracted from the fundus images using pretrained Efficientnet-b0 model. Efficientnet models developed by Tan et al. [22] have gained a lot of attention due to the incorporation of compound scaling techniques. It evenly scales width, depth, and resolution with a given set of scaling factors described by the following equations.

$$\text{depth: } d = \delta^\theta \tag{6.1}$$

$$\text{width: } \omega = \psi^\theta \tag{6.2}$$

$$\text{resolution: } r = \zeta^\theta \tag{6.3}$$

where θ is a compound coefficient and δ, ψ, and ζ are the scaling coefficients of each dimension fixed by grid search method by satisfying the constraints $\delta \cdot \psi^2 \cdot \zeta^2 \approx 2$ and $\delta \geq 1, \psi \geq 1, \zeta \geq 1$. Efficientnet model enhances the overall performance by balancing all network dimensions in relation to available resources. The feature extraction module of the Efficientnet-b0 includes seven mobile inverted bottleneck convolution (MIBConv) blocks, each with different settings. These blocks also include squeeze and excitation blocks along with swish activation. Table 6.2 represents the description of the model's layers.

Efficientnet-b0, like other existing models, is trained using over one million images from the ImageNet database. As a result, the generic features of the images are learned effectively by the model. Hence, fine-tuning the model using the transfer learning technique makes the training process easy, compared with naive training [23]. The model is customized by transferring the initial layers while modifying the final layers. Fig. 6.6 depicts the feature extraction using the customized Efficientnet-b0 model. The customized model is trained with generated fundus images. The discriminative features

Table 6.2   Layer details of Efficientnet-b0 model.

$$k3: 3 \times 3, k5: 5 \times 5, and\ k1: 1 \times 1$$

Stage	Operator	Resolution	No. of channels	No. of layers
1	Conv $k3$	224 × 224	32	1
2	MIBConv1, $k3$	112 × 112	16	1
3	MIBConv6, $k3$	112 × 112	24	2
4	MIBConv6, $k5$	56 × 56	40	2
5	MIBConv6, $k3$	28 × 28	80	3
6	MIBConv6, $k5$	14 × 14	112	3
7	MIBConv6, $k5$	14 × 14	192	4
8	MIBConv6, $k3$	7 × 7	320	1
9	Conv $k1$/pooling/FC	7 × 7	1280	1

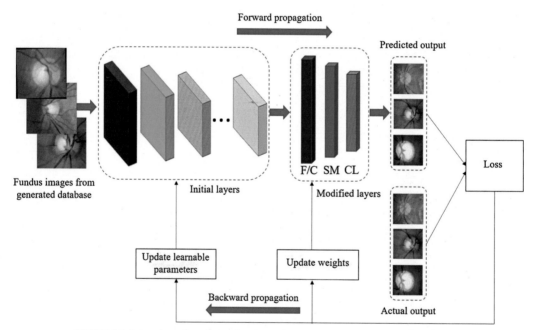

FIGURE 6.6 Extraction of pertinent features using customized Efficientnet-b0 model.

are extracted from the global average pooling layer (layer 287) of the trained network. These features indicate the discrimination between normal, early glaucoma, and advanced glaucoma images.

Feature selection: Selection of robust features is essential to categorize the stages of glaucoma. This is achieved by using the Haar wavelet transform and principal component analysis technique. Firstly, the Haar wavelet is used to perform single-level decomposition on the extracted features from the pretrained model. The decomposition provides an approximation band and detail band. The approximation band contributes toward glaucoma features more than the detail band. Thus, considering the approximation band alone may greatly reduce the number of features needed for classification [24]. Secondly, principal component analysis (PCA) technique is employed to generate the robust features from the decomposed bands. Reconstruction of the features is performed by zeroing the coefficients of the detail band and altering the coefficients of the approximation band. After reconstruction, PCA is applied to select the more robust features needed for further classification.

Classification: The type and nature of the classifier may greatly enhance the performance of glaucoma classification. In this work, the obtained robust features are categorized into three classes using Bi-LSTM network [25]. It is a combination of two LSTMs, where one LSTM takes the input sequence from start to end while the other takes the input sequence in the reversed order. The Bi-LSTM network classifies the features using the sequential arrangement of layers (such as input sequence, Bi-LSTM, fully connected, softmax, and classification).

6.4 Experimental findings

During training, the initial layers of Efficientnet-b0 model are retained, while the final layers are modified to the three class problem. The distribution of images in original and augmented databases is represented in Table 6.3. The model is trained with 11,442 advanced glaucoma images, 11,329 early glaucoma images, and 11,584 normal images. The learnable parameters are updated using the Adam optimizer [26], as it performs better for the majority of retinal databases [27]. A batch size of 32, maximum epochs of 20, and an initial learning rate of 0.0001 are used in this work. With these optimum

Table 6.3   Quantity of training and testing images in original and generated databases.

Database	Training images	Testing images
Original	GA: 327	GA: 140
	GE: 202	GE: 87
	N: 552	N: 236
Generated	GA: 11,442	GA: 4903
	GE: 11,329	GE: 4855
	N: 11,584	N: 4964

GA represents advanced glaucoma images, *N* indicates normal images, and GE indicates early glaucoma images.

values, the learnable parameters are updated to minimize the cross entropy loss defined as

$$\text{loss} = -\sum_{i=1}^{S}\sum_{j=1}^{C} t_{ij} ln\left(y_{ij}\right),$$

(6.4)

where S is the number of samples, C is the number of classes, t_{ij} indicates actual output, and p_{ij} represents the predicted output.

A total of 14,722 images (4903 advanced glaucoma images, 4855 early glaucoma images, and 4964 normal images), which are completely unknown to the customized model are used for testing. In this work, the model is tested in various configurations, which are listed in the following.

- C1: This configuration uses the probability scores obtained by softmax activation to classify the images into normal, early glaucoma, and advanced glaucoma.
- C2 and C3: Configuration C2 and C3 indicate the classification of extracted features using SVM and KNN classifier, respectively. The features needed for classification are obtained from the global average pooling layer of the trained model.
- C4: This configuration employs the Bi-LSTM network for classification. The model is trained and tested with the extracted features. Based on the obtained probability scores using softmax activation, classification is performed.
- C5, C6, and C7: These configurations investigate the effectiveness of feature selection using Haar wavelet transform toward the task of classifying glaucoma stages. The transformed features are classified using SVM, KNN, and Bi-LSTM in C4, C5, and C6 configurations, respectively.
- C8, C9, and C10: These configurations analyze the importance of robust features for glaucoma classification. The robust features are selected by applying principal component analysis. The configurations C8, C9, and C10 represent the classification of robust features using SVM, KNN, and Bi-LSTM, respectively.

The efficacy of the various configurations is evaluated using the following performance metrics:

6.4.1   Accuracy

Accuracy (ACC) is the standard metric that indicates the number of samples correctly classified by the model. It is given by

$$\text{ACC} = \frac{\text{TPA} + \text{TPE} + \text{TPN}}{N},$$

(6.5)

where TPA, TPE, and TPN represent the advanced glaucoma samples correctly classified, early glaucoma images correctly classified, and normal images correctly classified, respectively. N indicates the total number of samples.

6.4.2   Sensitivity

SEN_A, SEN_E, and SEN_N indicate the percentage of advanced glaucoma images, percentage of early glaucoma images, and percentage of normal images correctly identified by the model, respectively. It is computed using

$$SEN_A = \frac{TPA}{TPA + FNA},$$
6.6)

$$SEN_E = \frac{TPE}{TPE + FNE},$$
(6.7)

$$SEN_N = \frac{TPN}{TPN + FNN},$$
(6.8)

Experiment is performed with original and generated databases. The model has achieved an overall classification accuracy of 60.5% when it is trained and tested with original images. The model may fail to capture the relevant features, required to classify the stages of glaucoma. This could be owing to the fact that there is a data disparity between the classes. This necessitates a suitable strategy to enlarge the dataset. Hence, with oversampling data-level approach, a modified database is generated with a larger number of images. Classification of glaucoma stages is performed using various classifiers (such as softmax, SVM, KNN, and Bi-LSTM). The softmax classifier classifies the stages of glaucoma using the probability scores computed by the softmax activation function. This configuration correctly classifies 4547 advanced glaucoma images, 2723 early glaucoma images, and 4747 normal images. With this, an overall accuracy of 81.6% is obtained. Table 6.4 depicts the performance metrics obtained by various classifiers.

Classification is performed by the SVM classifier in configurations (such as C2, C5, and C8), whereas the KNN classifier is employed in various configurations like C3, C6, and C9, respectively. The SVM classifier seeks to find a hyperplane that maximizes the

Table 6.4   Performance metrics obtained by various configurations.

Performance metrics/Classifiers	ACC	SEN_A	SEN_E	SEN_N
C1 (Softmax)	81.6	92.7	56.1	95.6
C2 (SVM)	83	91.2	60.5	96.8
C3 (KNN)	81.5	94.9	52.2	97
C4 (Bi-LSTM)	83.4	85.6	65.2	99.2
C5 (Haar + SVM)	82.9	91.7	60.2	96.4
C6 (Haar + KNN)	81.6	95.3	52.3	96.8
C7 (Haar + Bi-LSTM)	83.5	93.5	63.8	96.2
C8 (Haar + PCA + SVM)	83.5	92.3	61.7	96.2
C9 (Haar + PCA + KNN)	82.3	89.7	60	97
C10 (Haar + PCA + Bi-LSTM)	84.65	93.5	63.85	96.2

separation of the data points into their respective classes. The main hyperparameter in SVM classifier is the kernel function. It is used to map the original dataset into a feature space so that the observations are more easily and accurately separable after this transformation. Here, a linear kernel function is employed to achieve better performance at a faster rate. The KNN classifier seeks to determine the degree of similarity between the test data and a specific number of neighboring data points in the training data. Euclidean kernel with K (number of neighbors) value of 5 is employed in KNN classifier for better classification of glaucoma and normal images. The features are extracted from the global average pooling layer (layer 287) of the trained network. With these 1280 features, the SVM and KNN models are trained and tested in C2 configuration and C3 configuration, respectively. It is inferred that the SVM classifier outperforms the KNN classifier, with an overall accuracy of 83%. A Bi-LSTM network is employed for final classification in configurations (such as C4, C7, and C10). An overall classification accuracy of 83.4% is obtained in the C4 configuration. The model classifies the advanced glaucoma images with 85.6%, early glaucoma images with 65.2%, and normal images with 99.2%. It is inferred that the majority of the early glaucoma images are classified as normal images.

The performance of the classifier may get improved by incorporating the feature selection technique. This work employs the single level Haar wavelet transform to select useful features from the approximation band for further classification. Thus, the relevant features from the approximation band with a dimension of 640 are selected. These selected features are trained and tested using SVM and KNN classifiers in C5 and C6 configurations, respectively. An overall classification accuracy of 82.9% and 81.6% are reported in C5 and C6 configurations, respectively. In the C7 configuration, the Bi-LSTM network is trained with the same 640 features. This configuration classifies the advanced glaucoma images with 93.5%, early glaucoma images with 63.8%, and normal images with 96.2%. It is observed that an improvement in classification accuracy is obtained in this configuration.

The configurations C8, C9, and C10 illustrate the effectiveness of the wavelet transform and PCA technique. By zeroing the detail coefficients, reconstruction is performed using only the approximation coefficients. The robust features are selected by applying the principal component technique. Thus, the features get reduced from 640 to 254, after applying the PCA. With these robust features, SVM and KNN classifiers are trained and tested in configurations C8 and C9, respectively. Overall classification accuracy of 83.5% and 82.3% is obtained in C8 and C9 configurations, respectively. Although there is a little improvement in the classification accuracy, the computational complexity is reduced in C8, and C9, compared with C2 and C3 configurations.

With 254 features, the Bi-LSTM model has achieved an overall classification accuracy of 84.65%. This configuration (C10) classifies the advanced glaucoma images with 93.5%, early glaucoma images with 63.85%, and normal images with 96.2%. From the findings, it is inferred that the classification accuracy is improved when the classifiers are trained and tested with selected features, compared to actual features extracted from layer 287 of the pretrained model.

To interpret how the model performs classification, activation maps are visualized using gradient weighted class activation mapping (Grad-CAM). It is an explainability technique that can be used to help understand the predictions made by CNN. It produces a heat map that highlights the important regions of an image by using the gradients of the classification score with respect to the final convolutional layer. Fig. 6.7 depicts the sample original images and the visualization maps obtained using Grad-CAM. The first row depicts the original advanced glaucoma images and their CAM. The second row indicates the original early glaucoma images with their CAM. The original normal images and their respective CAM is indicated in third row. The Grad-CAM maps indicate that the model is primarily focusing on the optic disc region in the fundus image for prediction. This is observed in all the classes.

Several other existing CNN models such as Alexnet [6], Googlenet [28], Shufflenet [29], Mobilenet-v2 [30], and Resnet-18 [31] are implemented. Table 6.5 indicates the performance metrics obtained by the CNN architectures. Transfer learning is employed to customize the pretrained models. Alexnet architecture comprises 25 layers yields an overall classification accuracy of 74.2%. It is inferred that this model fails to classify the early glaucoma images. With 144 layers, Googlenet extracts the discriminative features

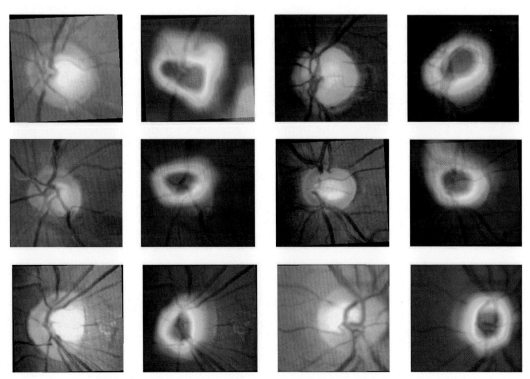

FIGURE 6.7 Sample fundus images and their corresponding class activation map.

Table 6.5 Performance metrics obtained by CNN architectures.

Performance metrics/CNN architectures	ACC	SEN$_A$	SEN$_E$	SEN$_N$
Alexnet	74.2	83.3	39.8	99
Googlenet	82	92	58.9	94.6
Shufflenet	81.5	95	54.8	94.2
Mobilenet-v2	80.5	92	51.8	97.3
Resnet-18	83.4	94.6	57.5	83.4
Proposed	84.65	93.5	63.85	96.2

from the images and achieves an accuracy of 82%. With 68 layers, Shufflenet achieves an overall accuracy of 81.5%. Mobilenet-v2 model comprises 155 layers and classifies the advanced glaucoma images with 92%, early glaucoma images with 51.8%, and normal images with 97.3%. By incorporating the residual connection, the Resnet-18 model outperforms the other models with an overall classification accuracy of 83.4%.

6.5 Conclusion

In this work, an unique approach for the classification of glaucoma stages using deep learning models is presented. The Efficientnet-b0 model is fine-tuned using a transfer learning approach. The features extracted from the trained model are classified using different classifiers. A total of nine classification approaches are experimented in this work. The efficacy of each configuration is evaluated on the database, comprising 49,077 fundus images (16,345 advanced glaucoma, 16,184 early glaucoma, and 16,548 normal images). It is inferred that robust features obtained using wavelet transform and PCA yields a better performance when classified using the Bi-LSTM network. This configuration correctly classifies the advanced glaucoma images with 93.5%, early glaucoma images with 63.85%, and normal images with 96.2%. It is inferred that integrating CNN and Bi-LSTM greatly enhances the performance metrics for the classification of glaucoma stages. The proposed approach may aid ophthalmologists in classifying the stages of glaucoma.

Conflicts of interest

The authors declare that they have no conflict of interest.

References

[1] A.K. Khurana, I. Khurana, Anatomy and Physiology of Eye, CBS Publishers and Distributors., 2006.
[2] Y.C. Tham, X. Li, T.Y. Wong, H.A. Quigley, T. Aung, C.Y. Cheng, Global prevalence of glaucoma and projections of glaucoma burden through 2040: a systematic review and meta-analysis, Ophthalmology 121 (11) (2014) 2081–2091.

[3] J. Mardeen, Glaucoma: the silent theft of sight, Nursing Times 110 (42) (2014) 20−22.

[4] P. Elangovan, R. Giritharan, M.K. Nath, O.P. Acharya, Review on localization of optic disc in retinal fundus images, in: 2018 International Conference on Applied Electromagnetics, Signal Processing and Communication (AESPC), IEEE, 2018, pp. 1−7.

[5] P. Elangovan, M.K. Nath, M. Mishra, Statistical parameters for glaucoma detection from color fundus images, Procedia Computer Science 171 (2020) 2675−2683.

[6] A. Krizhevsky, I. Sutskever, G.E. Hinton, Imagenet classification with deep convolutional neural networks, Neural Information Processing Systems 25 (2) (2012) 1097−1105.

[7] U. Raghavendra, H. Fujita, S.V. Bhandary, A. Gudigar, J. Hong Tan, U. Rajendra Acharya, Deep convolution neural network for accurate diagnosis of glaucoma using digital fundus images, Information Sciences 144 (29) (2018) 41−49.

[8] X. Chen, Y. Xu, D. Wong, T.-Y. Wong, J. Liu, Glaucoma detection based on deep convolutional neural network, Annual International Conference of the IEEE Engineering in Medicine and Biology Society (EMBC) (2015) 715−718. Milan, Italy, IEEE.

[9] P. Elangovan, M.K. Nath, Glaucoma assessment from color fundus images using convolutional neural network, International Journal of Imaging Systems and Technology 31 (2) (2021) 955−971.

[10] M.N. Bajwa, M.I. Malik, S.A. Siddiqui, A. Dengel, F. Shafait, W. Neumeier, S. Ahmed, Two-stage framework for optic disc localization and glaucoma classification in retinal fundus images using deep learning, BMC Medical Informatics and Decision Making 19 (1) (2019) 1472−6947.

[11] H. Fu, J. Cheng, Y. Xu, C. Zhang, D.W.K. Wong, J. Liu, X. Cao, Disc-aware ensemble network for glaucoma screening from fundus image, IEEE Transactions on Medical Imaging 37 (11) (2018) 2493−2501.

[12] M. Christopher, A. Belghith, et al., Performance of deep learning architectures and transfer learning for detecting glaucomatous optic neuropathy in fundus photographs, Scientific Reports 8 (2018) 1−13.

[13] S. Liu, S.L. Graham, A. Schulz, M. Kalloniatis, B. Zangerl, W. Cai, Y. Gao, B. Chua, H. Arvind, J. Grigg, D. Chu, A.K. A, Y. You, A deep learning-based algorithm identifies glaucomatous discs using monoscopic fundus photographs, Ophthalmology Glaucoma 1 (1) (2018) 15−22.

[14] H. Cho, Y.H. Hwang, J.K. Chung, et al., Deep learning ensemble method for classifying glaucoma stages using fundus photographs and convolutional neural networks, Current Eye Research 46 (10) (2021) 1516−1524.

[15] J. Gomez-Valverde, A. Anton, G. Fatti, B. Liefers, A. Herranz, A. Santos, C. Sánchez, M. Ledesma-Carbayo, Automatic glaucoma classification using color fundus images based on convolutional neural networks and transfer learning, British Journal of Ophthalmology 10 (2) (2019) 892−913.

[16] A. Diaz-Pinto, S. Morales, V. Naranjo, T. Kohler, J. Mossi, A. Navea, Cnns for automatic glaucoma assessment using fundus images:an extensive validation, BioMedical Engineering Online 18 (29) (2019) 1−19.

[17] Y. Zhen, L. Wang, H. Liu, J. Zhang, J. Pu, Performance assessment of the deep learning technologies in grading glaucoma severity, ArXiv 9 (4) (2018) 297−314.

[18] P. Elangovan, D. Vijayalakshmi, M.K. Nath, Detection of glaucoma from fundus images using pre-trained densenet201 model, Indian Journal of Radio and Space Physics 50 (1) (2021) 33−39.

[19] Y.C. Ko, S.Y. Wey, W.T. Chen, Y.F. Chang, M.J. Chen, S.H. Chiou, C.J. Liu, C.Y. Lee, Deep learning assisted detection of glaucomatous optic neuropathy and potential designs for a generalizable model, PLoS One 15 (5) (2020) 1−15.

[20] J.M. Ahn, S. Kim, K.S. Ahn, S.H. Cho, K.B. Lee, U.S. Kim, A deep learning model for the detection of both advanced and early glaucoma using fundus photography, Current Eye Research 13 (11) (2018) 1516−1524.

[21] A. Rehman, I.A. Taj, M. Sajid, K.S. Karimov, An ensemble framework based on deep cnns architecture for glaucoma classification using fundus photography, Mathematical Biosciences and Engineering 18 (5) (2021) 5321−5346.

[22] M. Tan, Q.V. Le, Efficientnet: rethinking model scaling for convolutional neural networks, CoRR abs/1905 (2019) 11946.

[23] P. Elangovan, M.K. Nath, En-ConvNet: A novel approach for glaucoma detection from color fundus images using ensemble of deep convolutional neural networks, International Journal of Imaging Systems and Technology (2022) 1−15.

[24] M.K. Gar, G. Ravichandran, P. Elangovan, M.K. Nath, Analysis of diagnostic features from fundus images using multiscale wavelet decomposition, ICIC Express Letters: Part B Applications 10 (2) (2019) 175−184.

[25] M. Schuster, K.K. Paliwal, Bidirectional recurrent neural networks, IEEE Transactions on Signal Processing 45 (11) (1997) 2673−2681.

[26] D.P. Kingma, J. Ba, Adam: a method for stochastic optimization, in: International Conference on Learning Representations, 2015, pp. 1−13. San Deigo, CA.

[27] P. Elangovan, M.K. Nath, Performance analysis of optimizers for glaucoma diagnosis from fundus images using transfer learning, Lecture Notes in Electrical Engineering 749 (2021) 507−518. Springer, Singapore.

[28] C. Szegedy, W. Liu, Y. Jia, P. Sermanet, S. Reed, D. Anguelov, D. Erhan, V. Vanhoucke, A. Rabinovich, Going deeper with convolutions, in: 2015 IEEE Conference on Computer Vision and Pattern Recognition (CVPR), IEEE, Boston, MA, USA, 2015, pp. 1−9.

[29] X. Zhang, X. Zhou, M. Lin, J. Sun, Shufflenet: an extremely efficient convolutional neural network for mobile devices, 2018 IEEE/CVF Conference on Computer Vision and Pattern Recognition (2018) 6848−6856.

[30] M. Sandler, A.G. Howard, M. Zhu, A. Zhmoginov, L. Chen, Inverted residuals and linear bottlenecks: mobile networks for classification, detection and segmentation, CoRR abs/1801 (2018) 04381.

[31] K. He, X. Zhang, S. Ren, J. Sun, Deep residual learning for image recognition, in: 2016 IEEE Conference on Computer Vision and Pattern Recognition (CVPR), IEEE, Las Vegas, USA, 2016, pp. 770−778.

Diagnosis of ophthalmic retinoblastoma tumors using 2.75D CNN segmentation technique

T. Jemima Jebaseeli[1], D. Jasmine David[2]

[1]*DEPARTMENT OF COMPUTER SCIENCE AND ENGINEERING, KARUNYA INSTITUTE OF TECHNOLOGY AND SCIENCES, COIMBATORE, TAMIL NADU, INDIA;* [2]*DEPARTMENT OF ELECTRONICS AND COMMUNICATION ENGINEERING, KARUNYA INSTITUTE OF TECHNOLOGY AND SCIENCES, COIMBATORE, TAMIL NADU, INDIA*

7.1 Introduction

Over the last 60 years, retinoblastoma has become more common and it is now seen in one out of every 15,000 newborns. Every year, 300 instances of retinoblastoma are diagnosed in the United States [1–3]. Retinoblastoma pathophysiology is the genetic component of the disease. The anthology has provided the groundwork for genetics. Retinoblastoma was one of the earliest cancers to be investigated in terms of genetics. Consider chromosome 13 (13q14), which will house the Rb gene. The Rb gene is also called as a tumor suppressor gene. It encodes the tumor suppressor protein, which inhibits cancer and controls uncontrollable neoplastic cell growth. At the cell cycle governor, Rb protein inhibits the cell cycle. It enables certain cell multiplication while preventing other instances from growing [4,5]. When this is destroyed, retinoblastoma develops and inactivation by mutations in both Rb gene alleles causes uncontrolled growth of tumor cells [6–9]. Therefore, retinoblastoma requires the inactivation of both alleles. It is divided into two categories: hereditary and nonhereditary sporadic. In the heritable sporadic retinoblastoma, one of the mutant genes will be passed down through the sperms from the parent and called as initial hit [10]. As a result, one faulty gene will be present in the zygote, and every child's cell has a copy of the defective gene, which causes the cells to proliferate. As a result of these second strikes, other normal alleles are mutated.

The first symptoms of retinoblastoma appear even before the baby is born. The infant's development after birth has the second impact. In the nonheritable retinoblastoma, the offspring's germ cells will be normal and the first hit occurs during the formative period, and the second hit happens later. As a result, the two alleles mutate, resulting in cancer. The first hit in a hereditary tumor is inherited during the genetic mutation, while the second hit is caused by the mutation [11–16]. There will be no change in the parents in sporadic or nonhereditary malignancies. Both targeted alterations, however, were acquired after delivery [17]. The following are the two fundamental key identifying elements of the two-hit hypothesis.

As shown in Fig. 7.1, hereditary retinoblastoma is always bilateral (B/L) and nonhereditary retinoblastoma is always unilateral (U/L). But there are some exceptions in some cases. Hereditary retinoblastoma manifests as multiple spots in the eye, whereas nonhereditary retinoblastoma manifests as a single spot [18]. Hereditary retinoblastoma patients are more likely to develop other types of cancer such as pinealoblastoma, or trilateral retinoblastoma. Osteosarcoma, soft tissue sarcoma, and melanoma are among the other cancers that fall into this category. In a nonheritable retinoblastoma, secondary tumors are unlikely to form. There will be two blastomas in both eyes in trilateral retinoblastoma. As a result, the patient will have B/L retinal blastoma as well as pinealoblastoma, cancer that is developing in the pineal gland.

The existing segmentation algorithms have several limitations that affect requirements, such as incorrect malignant identification due to a lack of separation between the malignant portions and the surrounding tissues. The lack of connection in the tumor structure is quite intricate. It was necessary to strengthen the nanoscopic contrast between the item's extractions. Cancer receding from its extensive path is sometimes inadequate to reveal the whole segmentation effect. The tumor cells' overall structure is reduced when the tracking mechanism is terminated too soon. The proposed technique is founded on this motive and skillfully finds the foremost solution of segments equal to the expected target. This approach identifies tumor cells based on abnormal features. As a result, tiny tumors with no discontinuities in their structure are appropriate for observation. The article has been organized as follows. Section 7.2 discusses reports and issues of lesion segmentation methodologies. Section 7.3 defines the datasets intended for experimentation. Section 7.4 displays the suggested lesion segmentation methodologies of the linear predictive decision median filter (LPDMF) and 2.75D convolutional

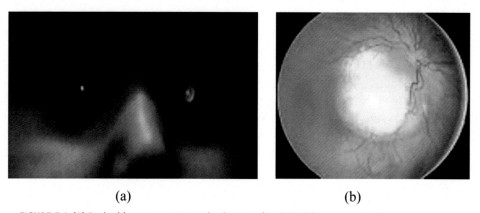

(a)   (b)

FIGURE 7.1 (A) Retinoblastoma cancer on both eyes of a child. (B) Tumor cells of a fundus image.

neural network. Section 7.5 depicts the findings and observations from the experiment. The paper concludes in Section 7.7.

7.2 Literature review

Retinoblastoma spreads in the eyeball and develops. It exits the retina in three ways: epiphytic, endogenous, and diffuse infiltration. Tumors develop in the vitreous space as much as many vitreous bacteria and spread into the subretinal space like subretinal bacteria. The tumor fills the vitreous space within the eyeball. It induces an exudative retinal detachment by forming subretinal fluid [19]. The tumor develops behind the subretinal region and frequently mixes. This point leads to extrinsic and intrinsic growth [20—22]. The rationalities are mirrored by diffuse infiltrative growth. It does not go backward or forwards, rather it traverses and spreads the retina on its own. It is a fast-growing tumor that spreads to the anterior chamber, forming a swarm of tumor seeds. It has no calcification and is easily overlooked.

7.2.1   Histopathology

Small primitive spherical cells with necrotic regions and isolated zones of dystrophic calcification make up affected retinoblastoma cells. The necrotic patches are pink, whereas the original spherical cells are blue. Pathological features of retinoblastoma include nesting of blue cells, a pink sea of necrosis, and calcification of rocks. Flexner-Wintersteiner rosette, Homer-Light rosette, and fluorite are the three most important histological findings, and biopsy is not indicated. Rosettes or garland chains occur in Flexner-Wintersteiner rosette tumors. There is an empty lumen in the center and it looks empty. Cytoplasmic elongation occurs centrally, and there is no basement membrane. Neutrophils are seen in the lumen of Homer's rosette. The rosettes will indicate well-differentiated tumors [23—25]. Fluorites are eosinophilic structures with bouquet-like features. It reflects the differentiation of photoreceptors. Retinocytoma is a tumor that solely contains laureates.

7.2.2   Expansion mode of retinoblastoma

If cancer spreads directly, it may infiltrate the optic node and therefore reach the central nervous system [26]. In addition, the tumor has progressed equally to the opposite sides of the optic nerve in the subarachnoid space. Because choroid and sclera in the orbit can develop due to direct growth, it is known as orbital retinoblastoma. Hematogenous or bleeding affects the liver, bones, and lungs, among other organs. The minor spread also occurs through the lymphatic system to the conjunctiva.

7.2.3   Segmentation techniques

The development of medical expert systems to automate diagnostic activities has received a lot of attention [27,28]. An expert system that gives precise responses based on predetermined rules. However, using static rules results in insufficient information and, as a result, no solution to new situations. By enhancing machine learning algorithms, the focus is on incorporating training data into machine learning. Through many studies, several researchers proposed ways for detecting retinoblastoma.

Carlos et al. [29] successfully segmented retinoblastoma cells using a CNN-based technique. The method does not make it easier to follow ocular cancers over time. As a result, the categorization findings caused the system to fail when assessing competing treatment plans. The complexity of the method is also raised as a result of the feature map created at each stage. In tumor classification, the algorithm relies entirely on expert annotated data. This may not be a suitable tool for rapid analysis of clinical diagnosis, as it is difficult to obtain real-time expert data for each clinical image. Some vessels are blended into the backdrop as a result of the preprocessing method [30]. The image is segmented using static thresholds and the output is supersaturated using the extreme learning machine approach. This indicates that this method might not be appropriate for clinical tumor categorization. In the computation of the extreme learning machine model used to create the output of the hidden layer, the computational matrix also depends on singular value decomposition. In multimodal pictures, this results in relatively poor efficiency when compared to higher order matrices [31]. Huu et al. [32] developed a 3T-MRI-based model that incorporates motion artifacts from the normal blink of the eye. Furthermore, tantalum staples lowered the model's accuracy. To address the drawbacks of existing techniques, well-defined parameters have been introduced to the proposed system's algorithms to increase clinical diagnostic and treatment accuracy.

To diagnose retinoblastoma in the past, images of the retinal background were required, which were separated from the background using image enhancement, segmentation, and sharp edge recognition. An algorithm known as apriori [33] is used for the resultant pixels. The fundus image was then analyzed using a Gaussian filter, a fast Fourier transform, and a logarithmic transform to diagnose retinoblastoma [34]. The intense learning technique [35] was created to address the drawbacks of feedforward artificial neural networks, particularly in terms of learning speed.

7.2.4   Clinical features

The most common 40% symptom in babies is leukocoria, which is characterized by a white eye reaction. When the eyes are lit with a lamp, the tumor reflexes [36]. The most prevalent retinoblastoma symptom that impacts vision is strabismus. It occurs at the tumor's macular location, leaving the child blind [37]. Furthermore, retinoblastoma can manifest in the cornea and is frequently misdiagnosed as megalocornea or congenital glaucoma. The reason is also blurred vision and red-eye. Retinoblastoma causes inflammatory reactions in the front region of the body and causes severe uveitis or acute red-eye disease. It masquerades as an ulcer, uveitis, or pus development. It results in retinoblastoma infiltrative spread. Proptosis symptoms emerge as a result of tumor necrosis caused by lid inflammation. As it progresses, it has an impact on the orbit, called orbital retinoblastoma, and the eyeball protrudes outward due to the functional orbital mass [38]. Advanced features of retinoblastoma are eyeball, congenital glaucoma, and orbital cellulitis tumor necrosis. Buphthalmos, such as retinoblastoma, is an excellent cornea imitator. Cellulitis is caused by an orbital retinoblastoma. A condition similar to hypopyon is diffuse infiltration. To detect retinoblastoma, a detailed diagnostic is necessary. Coats disease is a kind of retinoblastoma. It has a yellowish response called Xanthocoria, which has a red reflex, but retinoblastoma has a white reaction.

7.3 Materials

Screening neonates for retinoblastoma is critical for detecting early abnormalities. The assessment entails a thorough examination of the family's history. The full indirect ophthalmoscopy key examination is performed under anesthesia. Calcification is diagnosed with the ultrasonography (USG) B scan. It exhibits retinoblastoma-like hypertonic stippled patches. The examination method employs MRI, CT, and fundus imaging. The extraocular extension or orbital extension indicated by this scan has soft tissue involvement, according to MRI images. The hyperreflective and hyporeflective aspects of retinoblastoma can be seen on an MRI scan. The calcified area is apparent on the CT scan, and the tumor nucleates the eye. The Retcam pediatric camera is used to collect images of the patients' fundus. Individuals between the ages of 12 and 20 are included in the planned study. The average period between diagnosis and treatment is 7 days. The specialists who have been selected indicate the ground truth for review. In addition, images from the American Society of Retina specialists' public dataset are used for evaluation. For testing, the proposed approach consists of 240 pathogenic retinoblastoma images.

7.4 Methodology

The primary goal of this technique is to determine if each pixel in a fundus picture is a lesion or not. This paper proposes a new approach for identifying lesions in retinoblastoma pictures. It explains the originality of the proposed approach in the following way:

  i. The noisy pixels are replaced by the linear prediction filter and improve the similarity of the image's denoised pixels.
 ii. 2.75D CNN captures hierarchical texture details. This function achieves higher classification efficiency with fewer parameters.
iii. Creating patches from three separate perspectives lowers computational costs.
 iv. When compared to 3D CNN, 2.75D CNN with less parameter performs better.

   Fig. 7.2 depicts the proposed system's schematic architecture. To scan the retina, iPhone's fundus camera is used. The fundus image should be included to the Android

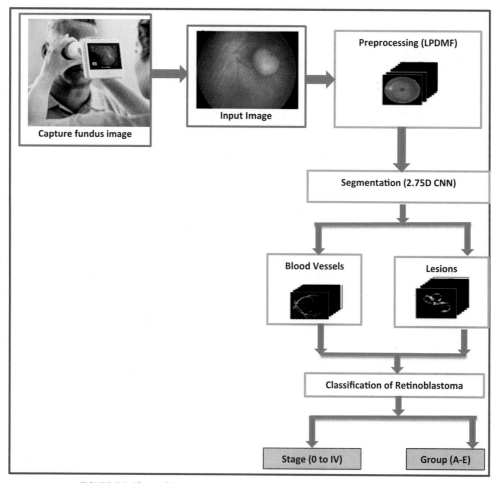

FIGURE 7.2 The architecture proposed for retinoblastoma cancer diagnosis.

application's GUI (Rbapp). Deep learning-based image processing technology is used in the suggested application. The retinal pictures must first be preprocessed using the LPDMF. The contrast and noise in the retinal images are improved using this procedure. Large blood vessels and malignant pixel strength values can be seen in the green channel of the retina. Green channel preprocessing is then performed on this image.

To handle real-time clinical images, an unsupervised segmentation approach is necessary. A 2.75D CNN-based deep learning model is used to segment and classify the pictures as blood vessels and lesions. The highlighted blood vessels in the algorithm are extracted, and the vascular map that results is used for blood vessel diagnostics. As a result of retinoblastoma, the blood vessel diameters will change. Based on the vascular width measurement, the approach can predict deviations from the conventional vessel width structure. The size of the blood vessels and the number of lesions seen in different quadrants of the images are used to estimate the severity of the illness. The severity of the sickness determines how retinoblastoma is graded and grouped into stages. Retinoblastoma is graded and classified into phases based on the severity of the illness. Murphree's worldwide retinoblastoma classification was utilized to classify the phases. The two aspects are staging and grouping. The child's survival is shown by staging, whereas the eye is saved by grouping. Levels 0 through IV are available. Stages 0 and 1 are not enucleated, but the tumor may be removed, stage 2 is enucleated but the tumor remains microscopic, stage 3 is regional expansion, and stage 4 is metastasis.

7.4.1 Preprocessing

The fundus camera photos revealed noise. It is added to the original image during capture or digital processing. It is difficult to spot them in fundus pictures because of the effect of noise on the microscopic tumor sizes. To maximize the impact of segmentation, images collected from diverse sources should be homogenized using preprocessing procedures. If the impulse noise has a high-density value, the median value of the substitute set will be calculated from the noise-free pixel values in the original set. The method enhances the correlation of the image's denoised pixels.

The input fundus image from the dataset being examined is shown in Fig. 7.3. The green channel of the image must first be extracted. LPDMF then preprocesses the image before it is utilized for segmentation and diagnosis.

7.4.2 2.75D CNN-deep learning model for segmentation

2.75D CNN is a multipath segmentation technique based on classification by voxel, where voxels represent the values of pixels in three-dimensional space. 2120 sample points were collected from one image. The sample is taken from the orthogonal direction of the plane. Fig. 7.4 shows that the system uses the 3D of the input image to create a patch. The 2D patch was created using a 2.75D CNN. The segmented images of the patch are combined using Bayesian aggregation techniques to produce the final segmented output.

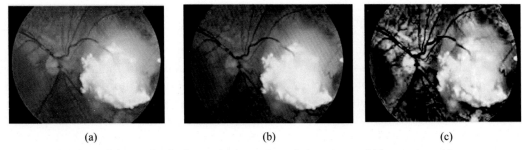

(a)   (b)   (c)

FIGURE 7.3 (A) Input fundus image, (B) green channel extraction, and (C) preprocessed image.

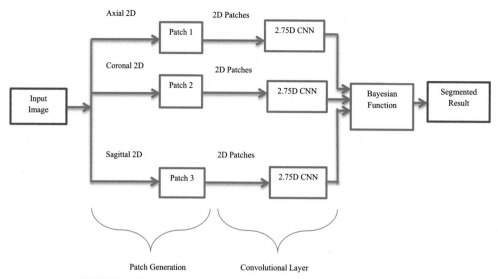

FIGURE 7.4 The architecture of 2.75D CNN model for tumor segmentation.

The proposed model provides the patch size in pixels, which contains all of the picture slice's information. The orthogonal planes of the CNN stack are the axial, sagittal, and coronal planes. Clinical data on lesions were acquired from all perspectives. As a consequence, the data are obtained from three adjacent image slices. A total of 32 patients were used to train the proposed model. During the training phase, each patch contains nine slices of the sample. In 2.75D, the suggested technique employs a new multichannel input approach. It analyses pixel spatial characteristics while sampling pixels in a 2D plane, a 3D CNN is used. This decreases the computational expenses of the method. Over 2D and 2.5D CNN, 2.75D CNN offers an advantage. To obtain high accuracy, there are fewer training input parameters and samples required. There were negligible data losses when obtaining information using such procedures. As a result, 2.75D is recommended to increase classification accuracy. The number of parallel CNN processes in this example, the spatial texture information from too many patches from each slice has grown.

7.5 Experimental analysis

Python is used to create the proposed model. The tests are run on a GNU/LINUX machine, and the approach is quicker to compute. The structures of the acquired tumor retinoblastoma demonstrate that the proposed technique is more effective when compared to the ground truth. The tumors that overlap are categorized and tracked back to the source node. The structure of the tumor structure is comparable to the ground truth image structure. The experimental outcome is shown in Fig. 7.5. In the initial

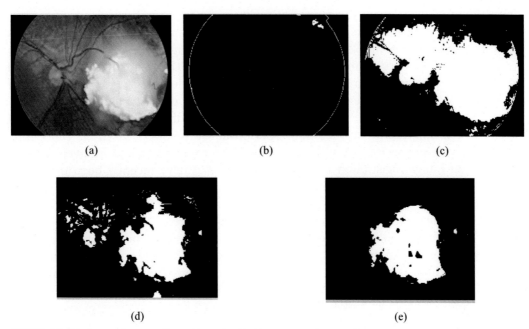

FIGURE 7.5 (A) Source image, (B) apply mask, (C–E) segmentation result at i = 1, i = 2, i = 3, and (E) final segmented result.

version, the optical disc is removed so that the malignancy may be traced correctly from its root node without being mistaken. The optic disc is divided in the first iteration to prevent cancer cells from taking the shortest path across the optic disc. In the following iteration, the discovered tumor cells are connected to their nearest neighbors. The picture is segmented into three rounds using the 2.75D CNN model. The morphological process begins at the top of the picture and continues throughout. The network sends the pixels that are relevant to the pixels of the next neighbor. The pixels in the front, which are unaffected by the pixels at the end of the rows, have their borders erased.

The proposed technique is sufficiently thorough to detect a single kind of cancer with no variation. The technique is sensitive in detecting nonmalignant pixels. It is also possible to separate tumor pixels from regular pixels. The proposed method has an average sensitivity of 98.96%, specificity of 99.32%, and accuracy of 99.82%. The average processing time for a single picture is 0.66 s. Even on high-resolution images, the technique decreases processing time.

7.6 Discussions

The proposed technique has achieved maximum efficiency; there are still a few issues that need to be addressed. The resulting output differs greatly from the ground-truth depiction of small tumor pixels. As a result, make use of a large number of specialists who will manually segment the photos of reality to be created. In such a setting, there will be better certainty around the tiniest cancerous sites and magnitudes. Light or dark areas in pathological photos do not affect the proposed approach. There have been few automated strategies for segmenting MRI ocular cancers. The first studies concentrated on segmenting retinoblastoma in children. For two deep learning methods, a 3D-Unet [18] and a 3D-CNN [39] concentrated on extremely small datasets of 16 and 32 retino-blastoma eyes, respectively. However, with an average dice similarity coefficient of roughly 62%, the tumor segmentation performance reported in these pioneering research initiatives was low. Only four instances have been studied qualitatively, despite the fact that image registration and the MRI threshold have recently been utilized to analyze retinoblastoma tumors.

The therapy of retinoblastoma is determined by grouping. From A through E, there are five groupings. Group A has a very low likelihood, Group B has a low probability, Group C has a low probability, Group D has a high probability, and Group E has a very high probability. The tumor in Group A is less than 3 mm in size, and there is no foveal subretinal seeding. The tumor size in Group B could be anywhere, but there is no seed. Group C tumors are confined or have localized and vitreous seeding. At this point, the tumor nipple cell has been broken down and dispersed throughout the vitreous cavity. In group D, there was widespread vitreous seeding and numerous dispersed tumor cells that migrated into the vitreous cavity. Group E is a large and sophisticated stage. The tumor is diffusely infiltrating at this stage and expands as an intra-ocular bleeding type, when it comes into contact with the lens, it might cause neovascular glaucoma.

Retinoblastoma is classified as an "H" cancer by the American Joint Cancer Committee. The first retinoblastoma malignancy was assigned this "H" status. The letter "H" stands for heritable. Hereditary cancer is extremely deadly [18]. The three pillars of therapy are to save lives, conserve the planet, and retain the minimum function of the vision. The four available therapy choices are as follows. Small cancers can be treated with laser or cryo-therapy as a focused treatment. The radiation shield is used to provide radiation to the tumor during local treatment. Chemotherapy is the most usually used treatment [39]. If the patient has widespread spreading, enucleate treatment is administered.

Chemotherapy is used when there is intraocular or extraocular retinoblastoma with local or regional dissemination. If focused or local therapy fails to heal the tumor, chemotherapy is administered and has a high success rate. Chemotherapy is most usually administered intravenously, intraarterially, or intravitreally. The tumor shrinks as a result of this treatment. The earliest way of preserving the life of a statement of Group E circumstances is enucleation. For as long as possible, the eyeball and optic nerve will be removed. Malignancies will not spread as a result. A 19–20 mm stump of the optic nerve is severed and removed during this treatment, when no other options are available. Myoconjunctival technique is the name for this approach. The optic nerve and eye are gone, but the conjunctiva is intact, the muscular tenons are intact, and the conjunctiva is intact. Muscle, tenons, and conjunctiva are also sutured to a synthetic implant. The gaps are held in place by the conformer.

7.7 Conclusion

In clinical research, it is critical to appropriately identify the malignancy. Several methods for detecting retinoblastoma tumors from the optic disc are ineffective. The proposed methodology's success measures have been updated by the result review. The proposed method removes background noise while subjectively distinguishing tumors. The findings show that the proposed approaches complement supervised methods well. In addition, the experts investigated the proposed segmentation results and determined that the quality remained intact. By evaluating the retina, the suggested approach simplifies the job of eye specialists in diagnosing retinoblastoma patients.

References

[1] M.A. Dyer, Dyer, Lessons from retinoblastoma: implications for cancer, development, evolution, and regenerative, Trends in Molecular Medicine 22 (10) (2016).

[2] A.H. Skalet, D.S. Gombos, B.L. Gallie, J.W. Kim, C.L. Shields, B.P. Marr, S.E. Plon, Screening children at risk for retinoblastoma—consensus report from the American association of ophthalmic oncologists and pathologists, Ophthalmology 125 (3) (2018).

[3] J.D. Fish, J.M. Lipton, P. Lanzkowsky, Lanzkowsky, Manual of Pediatric Hematology and Oncology, Academic Press, 2022, pp. 583—596.

[4] L. Sheng, J. Wu, X. Gong, D. Dong, X. Sun, SP1-induced upregulation of lncRNA PANDAR predicts adverse phenotypes in retinoblastoma and regulates cell growth and apoptosis in vitro and in vivo, Gene 668 (2018) 140—145.

[5] L. Lumbroso-Le Rouic, R. Blanc, C. Saint Martin, A. Savignoni, H.J. Brisse, N. Pierrat, C. Lévy-Gabriel, A. Matet, F. Doz, I. Aerts, N. Cassoux, Selective ophthalmic artery chemotherapy with melphalan in the management of unilateral retinoblastoma: a prospective study, Ophthalmology Retina 5 (8) (2021) 30—37.

[6] D. Essaid, A. Tfayli, P. Maillard, C. Sandt, V. Rosilio, A. Baillet-Guffroy, A. Kasselouria, Retinoblastoma membrane models and their interactions with porphyrin photosensitisers: an infrared microspectroscopy study, Chemistry and Physics of Lipids 215 (2018) 34—45.

[7] A.H. Skalet, D.S. Gombos, B.L. Gallie, J.W. Kim, C.L. Shields, B.P. Marr, S.E. Plon, P. Chévez-Barrios, Screening children at risk for retinoblastoma: consensus report from the American association of ophthalmic oncologists and pathologists, Ophthalmology 125 (3) (2018) 453—458.

[8] L.A. Dalvin, D. Ancona-Lezama, J.A. Lucio-Alvarez, B. Masoomian, P. Jabbour, C.L. Shields, Ophthalmic vascular events after primary unilateral intra-arterial chemotherapy for retinoblastoma in early and recent eras, Ophthalmology 125 (11) (2018) 1803—1811.

[9] A.S. Tomar, P.T. Finger, B. Gallie, T.T. Kivelä, A. Mallipatna, C. Zhang, J. Zhao, et al., High-risk pathology based on presenting features in advanced intraocular retinoblastoma: a multicenter, international data-sharing AJCC study, Ophthalmology 129 (8) (2022) 923—932.

[10] D. Jude, Anitha, S. Le, M. Mittal, Diabetic retinopathy diagnosis from retinal images using modified hopfield neural network, Journal of Medical Systems 247 (42) (2018).

[11] F.Z. Outtaleb, L. Kora, G. Jabrane, N. Serbati, L. El Maaloum, B. Allali, A. El Kettani, H. Dehb, 13q interstitial deletion in a Moroccan child with hereditary retinoblastoma and intellectual disability: a case report, Annals of Medicine and Surgery 60 (2020) 334—337.

[12] D.H. Abramson, J.H. Francis, I.J. Dunkel, B.P. Marr, S.E. Brodie, Y. Pierre Gobin, Ophthalmic artery chemosurgery for retinoblastoma prevents new intraocular tumors, Ophthalmology 120 (3) (2013) 560−565.

[13] T. Jemima Jebaseeli, C. Anand Deva Durai, J. Dinesh Peter, IOT based sustainable diabetic retinopathy diagnosis system, Sustainable Computing: Informatics and Systems 28 (2020).

[14] Ann Patrice Sheehan, Retinoblastoma: early diagnosis is crucial, Journal of Pediatric Health Care 34 (6) (2020) 601−605.

[15] F.Y. Lin, M.M. Chintagumpala, Neonatal retinoblastoma, Clinics in Perinatology 48 (1) (2021) 53−70.

[16] D.H. Abramson, I.J. Dunkel, S.E. Brodie, Y. Brian Marr, P. Gobin, Superselective ophthalmic artery chemotherapy as primary treatment for retinoblastoma (chemosurgery), Ophthalmology 117 (8) (2010) 1623−1629.

[17] D. Jude, Balas, Valentina, Anitha, Hybrid neuro-fuzzy approaches for abnormality detection in retinal images, management of retinal detachment in retinoblastoma with globe conserving treatment, Journal of Current Ophthalmology (2018) 1−6.

[18] A.N. Pandey, Retinoblastoma: an overview, Saudi Journal of Ophthalmology 28 (2014) 310−315.

[19] E.B. Turkoglua, M.K. Erol, Optical coherence tomography findings in a case with cavitary retinoblastoma Résultats de tomographie par cohérence optique dans un cas avec rétinoblastome cavitaire, Journal Français d'Ophtalmologie 44 (2) (2021) e97−e98.

[20] D. Jude, O. Deperlioglu, U. Kose, An enhanced diabetic retinopathy detection and classification approach using deep convolutional neural network, Neural Computing & Applications 32 (2020) 707−721.

[21] T. Jemima Jebaseeli, C. Anand Deva Durai, J. Dinesh Peter, Segmentation of retinal blood vessels from ophthalmologic diabetic retinopathy images, Computers & Electrical Engineering 73 (2019) 245−258.

[22] S. Suzuki, T. Yamane, M. Mohri, A. Kaneko, Selective ophthalmic arterial injection therapy for intraocular retinoblastoma: the long-term prognosis, Ophthalmology 118 (10) (2011) 2081−2087.

[23] M.L. Huang, H.Y. Chen, Development and comparison of automated classifiers for glaucoma diagnosis using Stratus optical coherence tomography, Investigative Ophthalmology & Visual Science 46 (2005) 4121−4129.

[24] J.P. Fernandez, A.A. Haider, L. Vajzovi, A. Ponugoti, M.P. Kelly, M.A. Materin, Optical coherence tomography angiography microvascular variations in pre- and posttreatment of retinoblastoma tumors, Ocular Oncology and Pathology 7 (2021) 330−339.

[25] Y. Luo, C. Zhou, F. He, J. Fan, X. Wen, Y. Ding, Y. Han, J. Ding, et al., Contemporary update of retinoblastoma in China: three-decade changes in epidemiology, clinical features, treatments, and outcomes, American Journal of Ophthalmology 236 (2022) 193−203.

[26] C. Ciller, S.I.S. De Zanet, S. Apostolopoulos, F.L. Munier, S. Wolf, J.-P. Thiran, M.B. Cuadra, R. Sznitman, Automatic segmentation of retinoblastoma in fundus image photography using convolutional neural networks, Investigative Ophthalmology & Visual Science 58 (8) (2017).

[27] C. Ciller, S. De Zanet, K. Kamnitsas, P. Maeder, B. Glocker, F.L. Munier, D. Rueckert, J.-P. Thiran, M. B. Cuadra, R. Sznitman, Multi-channel MRI segmentation of eye structures and tumors using patient-specific features, PLoS One 12 (3) (2017).

[28] W.J. Muen, J.E. Kingston, F. Robertson, S. Brew, M.S. Sagoo, M. Ashwin Reddy, Efficacy and complications of super-selective intra-ophthalmic artery melphalan for the treatment of refractory retinoblastoma, Ophthalmology 119 (3) (2012) 611−616.

[29] Machine, I. Jaya, U. Andayani, B. Siregar, T. Febri, D. Arisandi, Identification of retinoblastoma using the extreme learning, Journal of Physics 1235 (1) (2019).

[30] H.-G. Nguyen, R. Sznitman, P. Maeder, A. Schalenbourg, M. Peroni, J. Hrbacek, D.C. Weber, A. Pica, M.B. Cuadra, Personalized anatomic eye model from T1-weighted volume interpolated gradient echo magnetic resonance imaging of patients with uveal melanoma, International Journal of Radiation Oncology, Biology, Physics 102 (4) (2018) 813−820.

[31] P.K.G. Kumzar, K. Kranukara, G.S. Thyagraju, An approach to the detection of retinoblastoma based on apriori algorithm, International Journal on Recent and Innovation Trends in Computing and Communication 5 (6) (2017) 733−738.

[32] C.K. Balasundari, L. Ulagammal, J. Sivapriya, Diagnosis retinal disease using image processing techniques, International Journal of Innovative Research in Computer and Communication Engineering 4 (2016).

[33] G.-B. Huang, Q.-Y. Zhu, C.-K. Siew, Extreme learning machine: theory and applications, Neurocomputing 70 (1−3) (2006) 489−501.

[34] A.W. Stacey, R. Bowman, A. Foster, et al., Incidence of retinoblastoma has increased: results from 40 European countries, Ophthalmology 128 (9) (2021) 1369−1371.

[35] A.L. Blitzer, S.A. Schechet, H.A. Shah, M.P. Blair, Retinoblastoma presenting as pseudohypopyon and preserved visual acuity, American Journal of Ophthalmology Case Reports 23 (2021).

[36] K. Chen, L.F. Goncalves, Aparna Ramasubramanian, prenatal diagnosis of retinoblastoma, Advances in Ophthalmology and Optometry 6 (2021) 55−68.

[37] S.J. Langenegger, S.E. Soliman, B.L. Gallie, Retinal mapping of heritable retinoblastoma, Journal of American Association for Pediatric Ophthalmology and Strabismus 23 (4) (2019) e38.

[38] D.H. Abramson, Y. Pierre Gobin, I.J. Dunkel, J.H. Francis, Successful treatment of massive choroidal invasion in retinoblastoma with intraarterial chemotherapy (ophthalmic artery chemosurgery), Opthalmology Retina 5 (9) (2020) 936−939.

[39] D. Jude, Rajinikanth, Venkatesan, V. Rao, S. Mishra, N. Hannon, R. Vijayarajan, S. Arunmozhi, Image fusion practice to improve the ischemic-stroke-lesion detection for efficient clinical decision making, Evolutionary Intelligence 14 (2) (2021) 1−11.

Fast bilateral filter with unsharp masking for the preprocessing of optical coherence tomography images—an aid for segmentation and classification

Ranjitha Rajan[1,2], S.N. Kumar[3]

[1]LINCOLN UNIVERSITY COLLEGE, KOTA BHARU, MALAYSIA; [2]LUC MRC, KUTTIKANAM, KERALA, INDIA; [3]DEPARTMENT OF EEE, AMAL JYOTHI COLLEGE OF ENGINEERING, KOTTAYAM, KERALA, INDIA

8.1 Introduction

Optical coherence tomography's (OCTs) real-time in vivo imaging makes it effective for diagnosing medical disorders [1]. OCT has a far higher spatial resolution than ultrasonic imaging [2]. OCT is a form of optical imaging that is identical to ultrasound but uses light instead of sound waves to produce images. Several noise mechanisms, primarily shot and speckle, impact OCT imaging. In Fourier domain OCT, shot noise is a fundamental limiting noise process that is characterized as an additive, uncorrelated Gaussian white noise (FD-OCT). Low-coherence interferometry is used in OCT to give high-resolution cross-sectional imaging [3]. The temporal and spatial coherence of backscattered optical light from tissue provides the basis for interferometry [4] in OCT imaging.

The OCT images in general are corrupted by speckle noise and are modeled as follows:

$$Nr = S \times Ns + Nb$$

where S is the desired noise-free image, and Ns and Nb are the speckle noise and the background noise, respectively [5].

Photon shot noise and photon excess noise are the most common types of light source noise. Photon shot noise is caused by the quantum nature of the source. Bose–Einstein statistics are used in OCT, super luminescent diodes, quantum well devices (below the lasing threshold), and wideband lasers, whereas Poisson statistics are used in almost monochromatic lasers [6,7]. Preprocessing is the initial stage in image processing application, and the outcome of preprocessing paves a way toward robust and efficient segmentation and classification [8,9]. The generative adversarial neural network was employed for the filtering of OCT images; apart from filtering, enhancement was also carried out that favors efficient classification results [10]. The boundary segmentation model was utilized in Ref. [11] for the extraction of retina from OCT images; before segmentation, the preprocessing was done by structural interpolation and lateral filtering [11]. The ensemble block matching 3D filter was utilized in Ref. [12] for the denoising of retinal fundus images, and proficient results were generated for the images from the public databases. Multiinput fully convolution neural network was employed in Ref. [13] for the filtering of OCT images, and efficient results were produced, when compared with the classical filtering algorithms. Clustering-based despeckling algorithm was proposed in Ref. [14] for the OCT images, and the K means clustering was utilized in this research work with a hybrid filtering approach comprising lee filter and adaptive wiener filter. Multilayer perceptron-based intelligent filtering algorithm was put forward in Ref. [15] for the filtering of OCT images. Adaptive total variation model was proposed in Ref. [16] for the filtering of OCT images, and the results outperform the median filtering algorithm. Neural network-based algorithms play a vital role in retinal disease diagnosis [17]. Modified Hopfield neural network was used in Ref. [18] for the diagnosis of diabetic retinopathy in retinal images, whereas in deep convolution neural network was utilized in Ref. [19] for enhanced diabetic retinopathy detection. Deep learning-based models gain prominence in medical image processing for disease diagnosis and prediction [20,21]. Section 2 focuses on the fast bilateral filter algorithm with an unsharp masking approach, and the results and discussion are highlighted in Section 3. The conclusion is finally drawn in Section 4.

8.2 Methodology

The mask (template or kernel) is applied to the image from left to right and top to bottom in the spatial domain filtering approach. The mask's center coefficient is placed in the pixel to be adjusted, and the filtering process is performed using four or eight neighborhood connectivity. The bilateral filter was found to be efficient for the filtering of retinal images [22], and nonlocal mean filter with double Gaussian function was employed for the filtering of OCT images [23]. The fast bilateral filter with an unsharp masking approach was proposed in this research work. The mathematical model of the filtering used in this research work is as follows

$$I_{\text{filtered}}(x,y) = I_{\text{input}}(x,y) + k\left[I_{\text{edges}}(x,y)\right]$$

where

$$I_{\text{edges}}(x,y) = I_{\text{input}}(x,y) - I_{\text{filtered}}(x,y)$$

In the classical unsharp masking filtering approach, the denoised version of the input image was generated from the Gaussian filter. The constant "k" is the scaling factor that attributes the effect of sharpening. The objective of the filtering algorithm is to minimize the effect of noise in image, and the edge preservation is the unique feature of the bilateral filter. The bilateral filter is a convolutional filter that may be used to denoise CT/ MR images in the spatial domain while preserving edge information. Based on the position of pixels and their gray level values, the bilateral filter mask coefficients are found. The geometric closeness and gray level similarity with the pixel in the bilateral filter window determine the bilateral filter coefficients for each pixel.

The bilateral filter kernel is the product of two subkernels: gray-level kernel (W_{gk}) and distance kernel (W_{sk}). The gray level kernel and the distance kernel are functions of gray level distance and spatial distance, respectively. The gray level kernel is defined as follows.

$$W_{gk} = \exp\left[\frac{-1}{2}\left(\frac{d_{gk}}{\sigma_{gk}}\right)^2\right]$$

where d_{gk} is the gray level distance and σ_{gk} is the distribution function for W_{gk}. The distance kernel is defined as follows.

$$W_{sk} = \exp\left[\frac{-1}{2}\left(\frac{d_{sk}}{\sigma_{dk}}\right)^2\right]$$

where d_{sk} is the spatial distance and σ_{dk} is the distribution function for W_{sk}. The kernel of the bilateral filter is represented as follows.

$$W_{bk} = W_{gk} \times W_{sk}$$

Smoothing is done in a bilateral filter when the pixels in the region are similar and is not being done when the pixels are distinct (possibility of edge). The bilateral filter is thus able to preserve the edges while reducing the noise by performing strong smoothing in the similar pixel region. The fast bilateral filter is utilized here for generating the filtered output, the fast bilateral filter is mathematically expressed as follows.

Consider a digital image {f(x):xεX}, the fast bilateral filter with Gaussian function is expressed as follows.

$$f_{BX(X)} = \frac{\sum_{j\varepsilon N} g\sigma_s(y)g\sigma_r(f(x-y) - f(x)f(x-y)}{\sum_{y\varepsilon N} g\sigma_s(y)g\sigma_r f(x-y) - f(x))}$$

In the above expression, $g_{\sigma s}$ and $g_{\sigma r}$ are the gaussian kernels

$$g_{\sigma s(X)} = \exp\left(\frac{-x^2}{2\sigma_{s^2}}\right) g_{\sigma r(y)} = \exp\left(\frac{-y^2}{2\sigma_{r^2}}\right)$$

Here N represents the neighborhood window size

The range kernel expression was approximated using polynomial and trigonometric functions $g_{\sigma r(y-\tau)=} \exp\left(\frac{-z^2}{2\sigma_{r2}}\right) \exp\left(\frac{-t^2}{2\sigma_{r2}}\right) \exp\left(\frac{-t\tau}{2\sigma_{r2}}\right)$

The expression comprises three terms: the first term is a scaling factor, the second term is a gaussian function centered at the origin, and the third term is a monotonically exponential function that will increase or decrease based on the value of z.

Applying Taylor series expansion to the third term

$$\exp\left(\frac{-t^2}{2\sigma_{r2}}\right) = \sum_{n=0}^{N} \frac{1}{n!} \frac{\tau t}{\sigma r^2} + \text{other terms of higher order}$$

The expression for the range kernel is written as follows:

$$g_{\sigma r(y-\tau)} = \exp\left(\frac{-z^2}{2\sigma_{r2}}\right) \exp\left(\frac{-t^2}{2\sigma_{r2}}\right) \sum_{n=0}^{N-1} \frac{1}{n!} \left(\frac{\tau t}{\sigma r^2}\right)^n$$

The above expression comprises a gaussian function and a polynomial of degree N. The gauss polynomial approximation of the fast bilateral filter is expressed as follows.

$$G_n(x) = \left(\frac{f(x)}{\sigma_r}\right)^n \text{ and } F_n(x) = \exp\left(\frac{-f(x)^2}{2\sigma_{r2}}\right) G_n(x)$$

The gaussian filtering of $F_n(x) = (F_n \times g_{\sigma s})(x) = \sum g_{\sigma s}(y) F_n(x-y)$

Replace $t = f(x-y)$ and $z = f(x)$ the gaussian polynomial approximation was utilized for

$$g_{\sigma r}(x-z) = \exp\left(\frac{-f(x)^2}{2\sigma_{r2}}\right) P(x)$$

The fast bilateral filter coupled with the unsharp masking approach generates proficient filtering results for optical coherence tomography images. The median filter is used in many applications; however, it will alter the nonnoisy pixels also. The gaussian filter in many cases will blur the output, and the edge preservation is poor in the classical bilateral filter. This research work couples the fast bilateral filter with the unsharp masking approach that efficiently filters the noise and preserves the edges too.

8.3 Results and discussion

The algorithms are developed in MATLAB 2017 a and tested on optical coherence tomography images from the database (https://www.kaggle.com/paultimothymooney/kermany2018). Images from four data sets are utilized in this research work for the validation of preprocessing algorithms. The details of the data set are depicted in Fig. 8.1. The data set comprises four classes of images.

The representative images from each type are taken for the analysis of filtering algorithms. In Fig. 8.2, (A,D) depict the input OCT image (O1) and its histogram, (B,E) bilateral filter output corresponding to input O1 and its histogram, and (C,F) fast bilateral output corresponding to input O1 and its histogram.

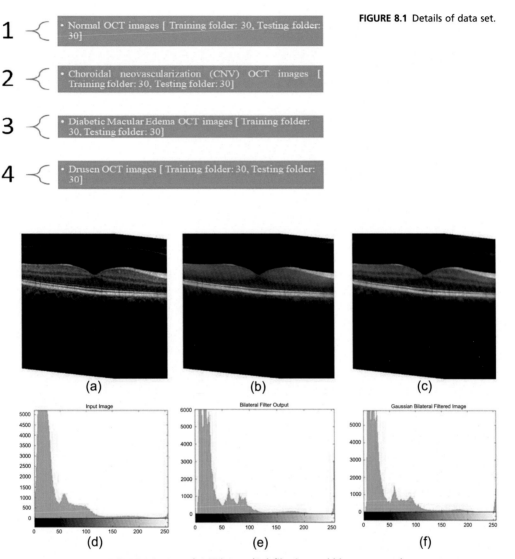

FIGURE 8.1 Details of data set.

1 — • Normal OCT images [Training folder: 30, Testing folder: 30]

2 — • Choroidal neovascularization (CNV) OCT images [Training folder: 30, Testing folder: 30]

3 — • Diabetic Macular Edema OCT images [Training folder: 30, Testing folder: 30]

4 — • Drusen OCT images [Training folder: 30, Testing folder: 30]

(a)   (b)   (c)

(d)   (e)   (f)

FIGURE 8.2 Normal OCT image (O1) filtering and histogram results.

In Fig. 8.3, (A) depicts the input OCT image (O2), (B) bilateral filter output corresponding to input O2, and (C) fast bilateral output corresponding to input O2.

In Fig. 8.4, (A) depicts the histogram of the input OCT image (O2), (B) histogram of bilateral filter output corresponding to input O2, and (C) histogram of fast bilateral output corresponding to input O2.

In Fig. 8.5, (A,D) depicts the input OCT image (O3) and its histogram, (B,E) bilateral filter output corresponding to input O3 and its histogram, and (C,F) fast bilateral output corresponding to input O3and its histogram.

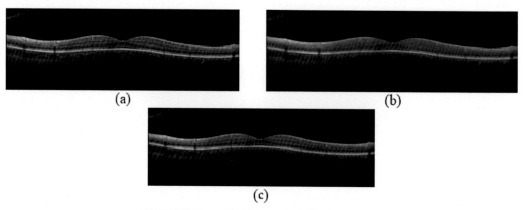

FIGURE 8.3 Normal OCT image (O2) filtering results.

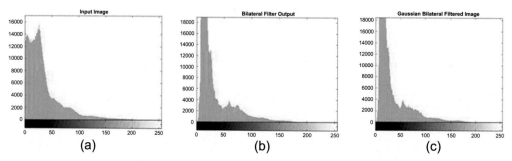

FIGURE 8.4 Normal OCT image (O2) histogram results.

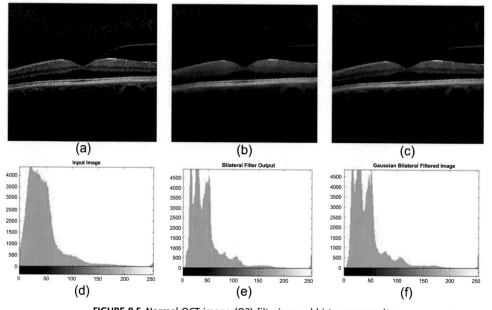

FIGURE 8.5 Normal OCT image (O3) filtering and histogram results.

In Fig. 8.6, (A,D) depicts the input OCT image (O4) and its histogram, (B,E) bilateral filter output corresponding to input O4 and its histogram, and (C,F) fast bilateral output corresponding to input O4 and its histogram.

In Fig. 8.7, (A) depicts the input OCT image (O5), (B) bilateral filter output corresponding to input O5, and (C) fast bilateral output corresponding to input O5.

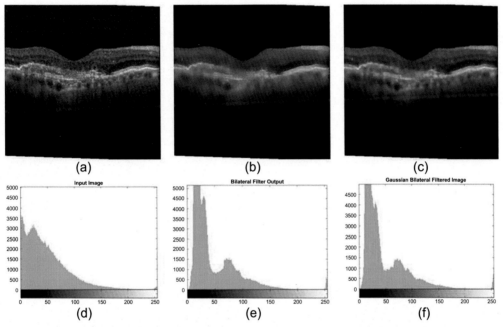

FIGURE 8.6 Choroidal neovascularization (CNV)OCT image (O4) filtering results.

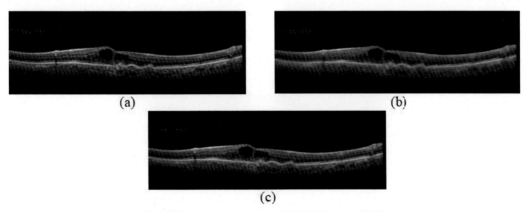

FIGURE 8.7 Choroidal neovascularization (CNV) OCT image (O5) filtering results.

In Fig. 8.8, (A) depicts the histogram of the input OCT image (O5), (B) histogram of bilateral filter output corresponding to input O5, and (C) histogram of fast bilateral output corresponding to input O5.

In Fig. 8.9, (A) depicts the input OCT image (O6), (B) bilateral filter output corresponding to input O6, and (C) fast bilateral output corresponding to input O6.

In Fig. 8.10, (A) depicts the histogram of the input OCT image (O6), (B) histogram of bilateral filter output corresponding to input O6, and (C) histogram of fast bilateral output corresponding to input O6.

In Fig. 8.11, (A,D) depicts the input OCT image (O7) and its histogram, (B,E) bilateral filter output corresponding to input O7 and its histogram, and (C,F) fast bilateral output corresponding to input O7 and its histogram.

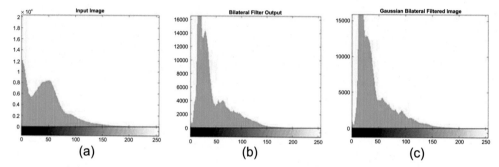

FIGURE 8.8 Choroidal neovascularization (CNV) OCT image (O5) histogram results.

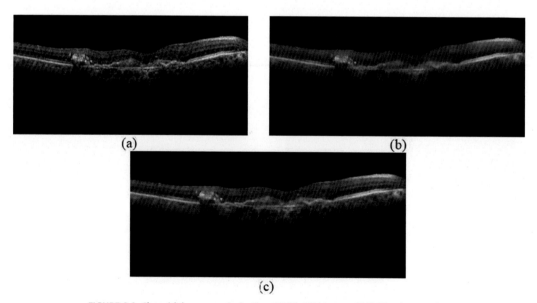

FIGURE 8.9 Choroidal neovascularization (CNV) OCT image (O6) filtering results.

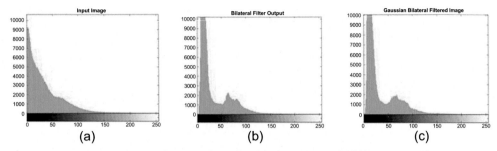

FIGURE 8.10 Choroidal neovascularization (CNV) OCT image (O6) histogram results.

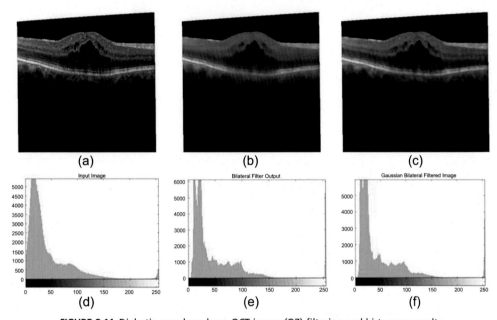

FIGURE 8.11 Diabetic macular edema OCT image (O7) filtering and histogram results.

In Fig. 8.12, (A) depicts the input OCT image (O8), (B) bilateral filter output corresponding to input O8, and (C) fast bilateral output corresponding to input O8.

In Fig. 8.13, (A) depicts the histogram of the input OCT image (O8), (B) histogram of bilateral filter output corresponding to input O8, and (C) histogram of fast bilateral output corresponding to input O8.

In Fig. 8.14, (A) depicts the input OCT image (O9), (B) bilateral filter output corresponding to input O9, and (C) fast bilateral output corresponding to input O9.

In Fig. 8.15, (A) depicts the histogram of the input OCT image (O9), (B) histogram of bilateral filter output corresponding to input O9, and (C) histogram of fast bilateral output corresponding to input O9.

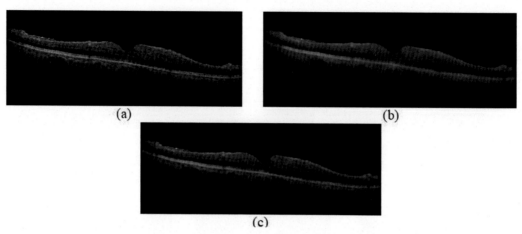

FIGURE 8.12 Diabetic macular edema OCT image (O8) filtering results.

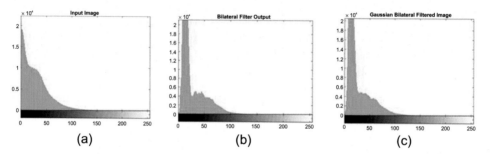

FIGURE 8.13 Diabetic macular edema OCT image (O8) histogram results.

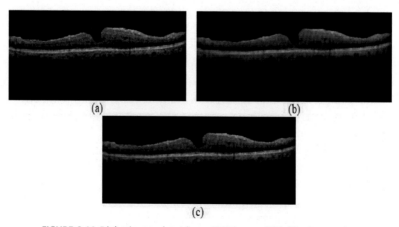

FIGURE 8.14 Diabetic macular edema OCT image (O9) filtering results.

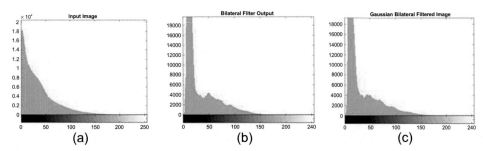

FIGURE 8.15 Diabetic macular edema OCT image (O9) histogram results.

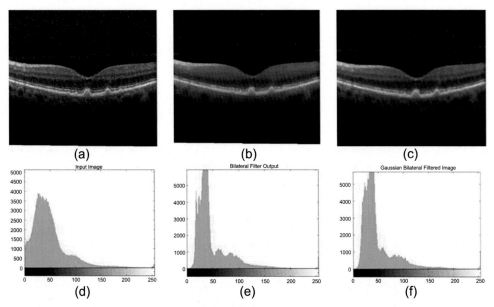

FIGURE 8.16 Drusen OCT image (O10) filtering and histogram results.

In Fig. 8.16, (A,D) depicts the input OCT image (O10) and its histogram, (B,E) bilateral filter output corresponding to input O10 and its histogram, and (C,F) fast bilateral output corresponding to input O10 and its histogram.

In Fig. 8.17, (A,D) depicts the input OCT image (O11) and its histogram, (B,E) bilateral filter output corresponding to input O11 and its histogram, and (C,F) fast bilateral output corresponding to input O11 and its histogram.

In Fig. 8.18, (A,D) depicts the input OCT image (O12) and its histogram, (B,E) bilateral filter output corresponding to input O12 and its histogram, and (C,F) fast bilateral output corresponding to input O12 and its histogram.

The performance of the filtering algorithms is validated by performance metrics and the following are used in this research work; peak to signal noise ratio (PSNR).

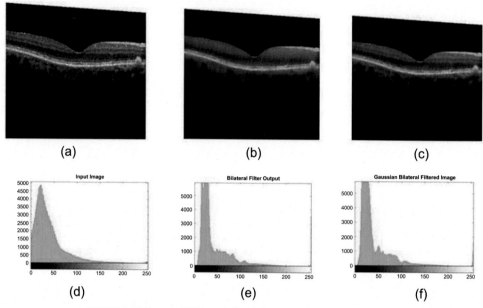

FIGURE 8.17 Drusen OCT image (O11) filtering and histogram results.

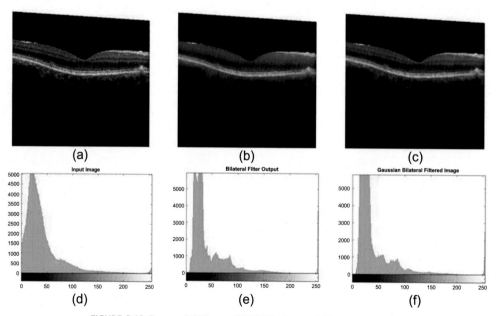

FIGURE 8.18 Drusen OCT image (O12) filtering and histogram results.

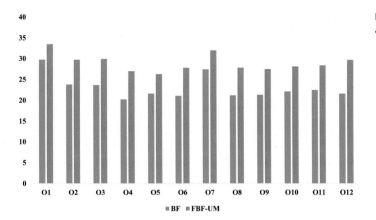

FIGURE 8.19 PSNR plot of filtering algorithms for OCT images.

The expression for PSNR and mean absolute error (MAE) is represented as follows in equations.

$$PSNR = 10 \, \log\left(\frac{255^2 x_0 x_1}{\|x_0 - x_1\|^2}\right) dB$$

$$MAE = \frac{\|x_0 - x_1\|}{x_0 x_1}$$

where x_0 and x_1 represent the original input image and the filtered image. The PSNR value will be high, and mean square error (MSE) and MAE values will be low for an efficient filtering algorithm. Fig. 8.19 depicts the PSNR plot of the filtering algorithms, and the fast bilateral filter with unsharp masking (FBF-UM) has high PSNR when compared with the bilateral filtering approach (BF).

The edges play a vital role in images, and for the validation of edges, Edge Preservation Index (EPI) is used [24]. The value of EPI is "1" for a good restoration algorithm, and a lower value indicates the discrepancy in the filtering approach.

$$EPI = \Gamma\left(\Delta I_1 - \frac{\overline{\Delta I_1}, \Delta I_2 - \overline{\Delta I_2})}{\sqrt{(\Delta I_1 - \overline{\Delta I_1}, \Delta I_1 - \overline{\Delta I_1})} \circ \Gamma\left(\Delta I_2 - \overline{\Delta I_2}, \Delta I_2 - \overline{\Delta I_2}\right)'}\right.$$

where I_1 is reference data, I_2 is measured data, N is the number of matrix elements, and $\overline{\Delta I}$ is the operation of Laplacian filtering in the region of interest. The edge preserving index of the FBF-UM approach was found to be better when compared with the BF approach in the filtering of OCT images (Table 8.1).

The Multiscale Structural Similarity Index (MSSIM) depicts the similarity between the input and the filtered image. The value of MSSIM, when closer to 1 represents the efficiency of restoration algorithms [25]. The MSSIM index is represented as follows.

Table 8.1 Edge preserving index values of filtering algorithms for OCT images.

Image details	Edge preservation index	
	FBF-UM	BF
O1	0.9665	0.9650
O2	0.6855	0.6504
O3	0.9060	0.9090
O4	0.7569	0.7649
O5	0.9502	0.9487
O6	0.7632	0.7422
O7	0.9480	0.9495
O8	0.8676	0.8505
O9	0.7487	0.7368
O10	0.8829	0.8796
O11	0.7703	0.7612
O12	0.8842	0.8817

Table 8.2 MSSIM Index values of filtering algorithms for OCT images.

Image details	MSSIM	
	FBF-UM	BF
O1	0.9642	0.9598
O2	0.8949	0.8885
O3	0.9143	0.9081
O4	0.8367	0.8304
O5	0.8313	0.8237
O6	0.8417	0.8355
O7	0.9435	0.9378
O8	0.8400	0.8327
O9	0.8489	0.8435
O10	0.8812	0.8746
O11	0.8726	0.8658
O12	0.9052	0.8959

$$\text{MSSIM} = \prod_{i=1}^{M} (\text{SSIM}_i)^{\beta_i}$$

where the value of β_i is estimated through psychophysical measurement.

$$\text{SSIM}_i = \begin{cases} \dfrac{1}{N_i}\sum C(t_{i,j}, s_{i,j})S(t_{i,j}, s_{i,j}); i = 1, 2, 3...M-1 \\ \dfrac{1}{N_i}\sum L(t_{i,j}, s_{i,j})C(t_{i,j}, s_{i,j})S(t_{i,j}, s_{i,j}); i = M \end{cases}$$

Table 8.3 FSIM index values of filtering algorithms for OCT images.

Image details	FSIM	
	FBF-UM	BF
O1	0.9195	0.8506
O2	0.8597	0.7633
O3	0.8547	0.7522
O4	0.8557	0.7548
O5	0.8590	0.7674
O6	0.8219	0.7084
O7	0.8988	0.8212
O8	0.8098	0.7030
O9	0.8207	0.7158
O10	0.8861	0.8045
O11	0.8751	0.7704
O12	0.8859	0.7909

The $t_{i,j}$, and s_{ij} represent j^{th} local image patches at the ith scale, N_i be the number of evaluation windows in the scale. The $L(t_{i,j}, s_{i,j})C(t_{i,j}, s_{i,j})$ and $S(t_{i,j}, s_{i,j})$ represents the luminance, contrast, and structural similarities (Table 8.2).

The MSSIM and FSIM index values are found to be better for the FBF-UM approach when compared with the BF approach for the filtering of OCT images. The feature similarity index (FSIM) also measures the similarity between the input and the filtered images. Closer the value of MSSIM and FSIM to 1, the better the efficiency of the filtering algorithm (Table 8.3).

8.4 Conclusion

This chapter proposes an efficient preprocessing approach for retinal optical coherence tomography images. The fast bilateral filter with an unsharp masking approach proposed in this research work produces proficient results when compared with the classical bilateral filtering approach. The performance metrics validation proves the proficiency of the filtering approach and the unsharp masking ensures edge preservation. The outcome of this research work paves a way toward the utilization of preprocessed OCT images for segmentation and classification. The future work is the development of deep learning-based classification model for retinal OCT images for disease diagnosis.

Acknowledgment

The authors would like to acknowledge the support from the Schmitt Center for Biomedical Instrumentation of Amal Jyothi College of Engineering for supporting this research work.

References

[1] D. Huang, et al., Optical coherence tomography, Science 254 (5035) (1991) 1178–1181.

[2] W. Drexler, J.G. Fujimoto (Eds.), Optical Coherence Tomography: Technology and Applications, vol 2, Springer, Berlin, 2015.

[3] Y. Ma, et al., Speckle noise reduction in optical coherence tomography images based on edge-sensitive cGAN, Biomedical Optics Express 9 (11) (2018) 5129–5146.

[4] S.K. Devalla, et al., A deep learning approach to denoise optical coherence tomography images of the optic nerve head, Scientific Reports 9 (1) (2019) 1–13.

[5] B. Qiu, et al., Noise reduction in optical coherence tomography images using a deep neural network with perceptually-sensitive loss function, Biomedical Optics Express 11 (2) (2020) 817–830.

[6] G. Brida, M. Genovese, I. RuoBerchera, Experimental realization of sub-shot-noise quantum imaging, Nature Photonics 4 (4) (2010) 227–230.

[7] M. Jensen, et al., Shot-noise Limited, Supercontinuum Based Optical Coherence Tomography, 2020 arXiv preprint arXiv:2010.05226.

[8] R.B. Jeyavathana, R. Balasubramanian, A. AnbarasaPandian, A survey: analysis on pre-processing and segmentation techniques for medical images, International Journal of Research and Scientific Innovation (IJRSI) 3 (2016).

[9] H. Hamid, NajmehSamadiani, Using morphologicaltransforms to enhance the contrast of medical images, TheEgyptian Journal of Radiology and Nuclear Medicine 46 (2) (June 2015) 481–489.

[10] Y. Huang, Z. Lu, Z. Shao, M. Ran, J. Zhou, L. Fang, Y. Zhang, Simultaneous denoising and super-resolution of optical coherence tomography images based on generative adversarial network, Optics Express 27 (9) (April 29, 2019) 12289–12307.

[11] Y. Ma, Y. Gao, Z. Li, A. Li, Y. Wang, J. Liu, Y. Yu, W. Shi, Z. Ma, Automated retinal layer segmentation on optical coherence tomography image by combination of structure interpolation and lateral mean filtering, Journal of Innovative Optical Health Sciences 14 (01) (January 3, 2021), 2140011.

[12] K. Naveed, F. Abdullah, H.A. Madni, M.A. Khan, T.M. Khan, S.S. Naqvi, Towards automated eye diagnosis: an improved retinal vessel segmentation framework using ensemble block matching 3D filter, Diagnostics 11 (1) (January 2021) 114.

[13] A. Abbasi, A. Monadjemi, L. Fang, H. Rabbani, Y. Zhang, Three-dimensional optical coherence tomography image denoising through multi-input fully-convolutional networks, Computers in Biology and Medicine 108 (May 1, 2019) 1–8.

[14] M.H. Eybposh, et al., Cluster-based filtering framework for speckle reduction in OCT images, Biomedical Optics Express 9 (12) (2018) 6359–6373.

[15] S. Adabi, S. Conforto, A. Clayton, A.G. Podoleanu, A. Hojjat, M. Avanaki, An intelligent speckle reduction algorithm for optical coherence tomography images, in: International Conference on Photonics, Optics and Laser Technology, PHOTOPTICS 2016, 2016, pp. 38–45.

[16] T. Wu, Y. Shi, Y. Liu, C. He, Speckle reduction in optical coherence tomography by adaptive total variation method, Journal of Modern Optics 62 (21) (2015) 1849–1855.

[17] J. Anitha, D. Selvathi, D.J. Hemanth, Neural computing based abnormality detection in retinal optical images, in: 2009 IEEE International Advance Computing Conference, IEEE, March 6, 2009, pp. 630–635.

[18] D.J. Hemanth, J. Anitha, L.H. Son, M. Mittal, Diabetic retinopathy diagnosis from retinal images using modified hopfield neural network, Journal of Medical Systems 42 (12) (December 2018) 1–6.

[19] D.J. Hemanth, O. Deperlioglu, U. Kose, An enhanced diabetic retinopathy detection and classification approach using deep convolutional neural network, Neural Computing & Applications 32 (3) (February 2020) 707–721.

[20] R. Jain, M. Gupta, S. Taneja, D.J. Hemanth, Deep learning based detection and analysis of COVID-19 on chest X-ray images, Applied Intelligence 51 (3) (March 2021) 1690–1700.

[21] R. Jain, N. Jain, A. Aggarwal, D.J. Hemanth, Convolutional neural network based Alzheimer's disease classification from magnetic resonance brain images, Cognitive Systems Research 57 (October 1, 2019) 147–159.

[22] Y. He, Y. Zheng, Y. Zhao, Y. Ren, J. Lian, J. Gee, Retinal image denoising via bilateral filter with a spatial kernel of optimally oriented line spread function, Computational and Mathematical Methods in Medicine 2017 (February 5, 2017).

[23] J. Aum, J.H. Kim, J. Jeong, Effective speckle noise suppression in optical coherence tomography images using nonlocal means denoising filter with double Gaussian anisotropic kernels, Applied Optics 54 (13) (May 1, 2015) D43–D50.

[24] J. Joseph, S. Jayaraman, R. Periyasamy, S. V Renuka, An edge preservation index for evaluating nonlinear spatial restoration in MR images, Current Medical Imaging 13 (1) (February 1, 2017) 58–65.

[25] U. Sara, M. Akter, M.S. Uddin, Image quality assessment through FSIM, SSIM, MSE and PSNR—a comparative study, Journal of Computer and Communications 7 (3) (March 4, 2019) 8–18.

Deep learning approaches for the retinal vasculature segmentation in fundus images

V. Sathananthavathi, G. Indumathi

MEPCO SCHLENK ENGINEERING COLLEGE, SIVAKASI, TAMIL NADU, INDIA

9.1 Introduction

Anatomical changes in the vascular pattern and variations in the proportion of artery and vein blood vessels are the usually observed symptoms of retinal diseases such as glaucoma, hypertension, and diabetic retinopathy and even for some cardiac-related diseases. Identification of retinal blood vessels in the retinal fundus images needs to overcome challenges like nonuniform contrast in the retinal fundus mages due to the capturing mechanism. The classification of the retinal blood vessels into arteries or veins also has challenging issues like uniform vessel thickness and brightness. Extraction of retinal blood vessels from fundus images and their classification into arteries or veins remains a challenging task. This challenge can be successfully addressed by the deep learning approaches. Deep learning has the capability to extract both low- and high-level features for segmenting anatomical structures like blood vessels from the fundus images.

In this chapter, fully convolved neural network architectures are discussed for the segmentation of retinal blood vessels and their classification into artery or vein vessels. The proposed deep neural network is the encoder–decoder architecture. The superiority of the proposed deep learning approaches compared to the supervised methodologies is also discussed. The improvement in the extraction of retinal blood vessels and their classification into arteries or veins are attempted through the dilated convolution on the proposed fully convolved neural network. From the experimentation, it is inferred that dilated convolution improves the segmentation of anatomical structures from the fundus images.

9.2 Significance of deep learning

The retinal vasculature segmentation and its classification into artery/vein have dominance by supervised methodologies. The unsupervised methodologies usually failed in efficient vasculature extraction due to the reasons like illumination and contrast are not uniform, undistinguished intensity range of vasculature, and other anatomical structures. Even supervised methods lead to unsupervised methodologies for both extraction of retinal blood vessel extraction and also its classification into artery or vein; still, there are adequate unfilled gaps by supervised methods exist. Handcrafted features fail in imitating the human neural scheme for differentiating vasculature from other regions. Deep learning approaches imitate the human way of extracting features for retinal vasculature extraction. The key advantages of deep learning approaches include the extraction of prominent features and eradication of the field expertise and selection of handcrafted features [1,2].

Convolutional neural network (CNN) [3,4] is precisely intended for processing pixel data, and it is the prominent architecture for deep learning. CNN-based supervised retinal vessel segmentation methodology is proposed Liskowski et al. [5]. Wang et al. applied CNN for extracting significant features and a random forest classifier for classifying blood vessels and other regions based on extracted features [6]. Zhou et al. applied CNN as a feature extractor with a dense conditional random field for retinal vessel segmentation [7].

Convolutional neural network-based networks are proposed for artery vein classification by Fantin et al. [8] and Weikala et al. [9]. A fully convolved neural network shows its significance by avoiding fully connected layers and hence reducing the network parameters. Methodologies based on unet or segnet come under fully convolve neural network architectures.

9.3 Convolutional neural network

In CNN, the arrangement of a neuron is similar to the visual stimuli by the frontal lobe, and it can associate subregions of the image. CNN can be able to observe both spatial and temporal dependencies on the image. Hence, it is possible to make the machine to see the details in an image similar to a human. Convolution operation preserves the relationship between the image pixels by learning image features from the receptive field of input data. Fig. 9.1 explains the convolution operation. To extract vital features from the image, different kernels are defined, and the features are learned. The result obtained by the convolutional layer is known as a feature map.

Striding and padding are the two parameters, which reign the size of the output image. The predefined step size followed by convolution operation horizontally is named as stride. When the striding value is "1" then the kernel will move one pixel at a time during the slide over the input image. When the stride is "2" then the filter will move by "2" pixels and so on. The process of adding zeros to control the output image size is the padding operation. The output image size from the convolutional layer is given by ((input image size − size of filter + 2 × padding)/stride) + 1.

The training of the CNN comprises the following phases: forward pass, loss computation, backward pass, and weight updating based on error. Forward pass is the estimation from the input layer to the output layer. Backward pass is the

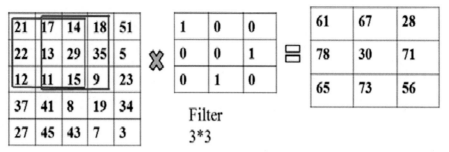

FIGURE 9.1 Convolution operation.

backpropagation from the output layer toward the input layer by recursively computing gradients. The loss computation is the mean square error computation between the target and the actual output achieved. Based on the loss, weight is updated by the weight update equation. A single iteration is the completion of all the above phases.

In the stochastic gradient descent, weight update after every sample leads to more oscillation and delays the convergence. In the batch gradient descent algorithm, weight is updated after the entire training samples are passed through the model. The requirement for memory increases as all the training samples need to be loaded. The mini batch gradient descent algorithm split the entire training samples into mini batches. With the intention of avoiding internal covariant shift by the mini batches, batch normalization is done and it is given by

$$BN_{\gamma,\beta} = \gamma \widehat{x}_i + \beta \tag{9.1}$$

where $\widehat{x}_i = \frac{x_i - \mu_B}{\sqrt{\sigma_B^2 + \varepsilon}}$, x_i is the values in the mini batch, μ_B is the batch mean, σ_B^2 is the batch variance, and ε is a constant added to the mini-batch variance for numerical stability. γ and β are the parameters to be learned to scale and shift the normalized value.

The pooling layer condenses features extracted by the convolutional layers. Two types of pooling are generally used: they are Max pooling and average pooling. In max pooling, the maximum among the specific region from the feature map is considered. In average pooling, the average of the specific region is considered. Fig. 9.2 shows the pooling operation for the filter of size (2 × 2) and stride of 2.

The activation function used is the Relu (rectified linear unit) function. It is widely accepted because of its superior performance over the Sigmoid function. As Sigmoid function has a very narrow nonzero derivative region, backpropagation is meaningless when the sigmoid function is in the saturation region. In the case of the Relu function, the derivative value is either one or zero. The computational cost is drastically reduced by the Relu function; hence, it is possible to reach the optimum solution even for the deep neural network. The Relu (rectified linear unit) activation function is described as

$$Relu(x) = \begin{cases} x, & |x \geq 0 \\ 0, & |x < 0 \end{cases} \tag{9.2}$$

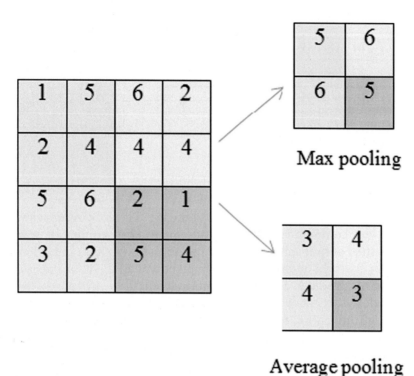

FIGURE 9.2 Pooling operation.

9.4 Fully convolved neural network

It is the convolutional network where all the learnable layers are convolutional layers and do not have any fully connected layers. Segnet architecture [10] is the popular fully convolved neural network for pixelwise semantic segmentation. Segnet architecture is the convolutional encoder–decoder architecture. The encoder stages of Segnet architecture are alike that of the convolutional layers in VGG16 [11]. Every encoder has its corresponding decoder that makes use of the pooling indices from the encoders. Due to these pooling indices sharing, the upsampling of the input on the decoder stages is in a nonlinear manner [12]. Segnet architecture comprises five stages of encoder and decoder. The first two encoder and decoder stages have two convolutional layers, whereas the deep three stages have three convolutional layers.

9.5 Retinal blood vessel extraction

9.5.1 Proposed fully convolved neural network

The proposed deep learning architecture comprises five stages of encoder–decoder, and all the stages are uniformly having two convolutional layers. In all the encoder stages, the

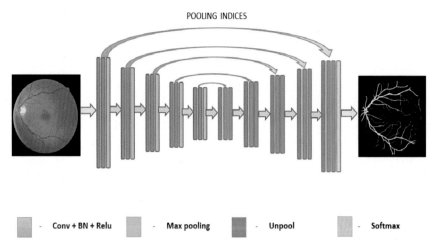

POOLING INDICES

- Conv + BN + Relu   - Max pooling   - Unpool   - Softmax

FIGURE 9.3 Fully convolved neural network for blood vessel extraction.

batch normalization layer and Relu layer follow each convolutional layer (Conv). Similar to Segnet architecture, pooling indices on the encoder side are shared on the decoder side. The final layer comprises the Softmax layer followed by the pixel classification layer. The Softmax layer gives probability value for the assigned classes, and the pixel classification layer labels every pixel in the image as a vessel or background (Fig. 9.3).

The encoder and decoder layer description of the proposed network is given in Tables 9.1 and 9.2, respectively. The kernel numbers are generally increased about the depth of the layers. With the aim of giving the same level of importance to low- and high-level features in the proposed work, convolutional layers in all the stages are having an equal number of kernels. The number of kernels in all the convolutional layers is set as 64. E_{ij} represents jth convolutional layer in the ith stage of the encoder, and D_{ij} represents jth convolutional layer in the ith stage of the decoder.

The weights of the proposed network are initialized with respect to Gaussian distribution with zero mean and 0.01 as standard deviation. Bias values are initially set as zero. The parameters of the network are updated based on the migration toward a negative gradient using the Gradient descent algorithm. Weights are updated based on the loss function minimization and are given by

$$\theta_{n+1} = \theta_n - \alpha \nabla E(\theta_n) + \gamma(\theta_n - \theta_{n-1}) \tag{9.3}$$

where "θ" is the parameter to be updated, α is the learning rate and it is set as 0.001, and γ is the momentum parameter and it is set as 0.9. ∇E is the gradient of the error function. l_2 regularization is also included in the loss function to avoid overfitting. Loss function with l_2 regularization is

$$E_r(\theta) = E(\theta) + \frac{\lambda w^T w}{2} \tag{9.4}$$

Table 9.1 Encoder layer description.

Layer name	Stride	Kernels	Padding	Filter size
E_{11}	1	64	1	(3,3)
E_{12}	1	64	1	(3,3)
Max pooling	2	64	0	(2,2)
E_{21}	1	64	1	(3,3)
E_{22}	1	64	1	(3,3)
Max pooling	2	64	0	(2,2)
E_{31}	1	64	1	(3,3)
E_{32}	1	64	1	(3,3)
Max pooling	2	64	0	(2,2)
E_{41}	1	64	1	(3,3)
E_{41}	1	64	1	(3,3)
Max pooling	2	64	0	(2,2)
E_{51}	1	64	1	(3,3)
E_{52}	1	64	1	(3,3)
Max pooling	2	64	0	(2,2)

Table 9.2 Decoder layer description.

Layer name	Kernels	Stride	Padding	Filter size
Unpool	—	—	—	—
D_{51}	64	1	1	(3,3)
D_{52}	64	1	1	(3,3)
Unpool	—	—	—	—
D_{41}	64	1	1	(3,3)
D_{42}	64	1	1	(3,3)
Unpool	—	—	—	—
D_{31}	64	1	1	(3,3)
D_{32}	64	1	1	(3,3)
Unpool	—	—	—	—
D_{21}	64	1	1	(3,3)
D_{22}	64	1	1	(3,3)
Unpool	—	—	—	—
D_{11}	64	1	1	(3,3)
D_{12}	64	1	1	(3,3)

where "λ" is the regularization factor of the weight parameter and it is assigned as 0.0001. The mini batch size is set as 128. Data augmentation is included to increase the number of training samples in diverse manner. Flipping and rotation are the operations included under data augmentation for the proposed work. Individual class probabilities can be computed using the Softmax function. The softmax function of an individual class is computed as

$$f(s_i) = \frac{e^{s_i}}{\sum_j^c e^{s_j}} \qquad (9.5)$$

The softmax activation function of class "s_i" is calculated using the scores gained by all classes "s_j." The frequency of occurrence of vessel pixels is less compared to that of nonvessel pixels and this leads to class imbalance problem. To overcome the class imbalance problem, class weightage is assigned based on the class occurrence frequency. The number of classes in the image is represented as "N" whose corresponding weights are denoted as $w = w^{(1)}, \dots, w^{(N)}$. The total class weight ($T(W)$) is given by

$$T(W) = \sum_{i=1}^N w^{(i)}. \qquad (9.6)$$

Class frequency is the ratio of each class weight to total class weight. Class frequency F is given by

$$F^{(i)} = \frac{w^{(i)}}{T(w)} \, i = 1, \dots, N \qquad (9.7)$$

Inverse class weight is the reciprocal of the class frequency. Inverse class weight $\left(w_I^{(i)}\right)$ is given by the following equation.

$$w_I^{(i)} = \frac{1}{F^{(i)}} \, i = 1, \dots, N \qquad (9.8)$$

The cross-entropy loss function is represented by

$$CE = -\sum_i^N t_i \log(s_i) \qquad (9.9)$$

where "t" represents the target and "s" represents the prediction. On applying cross-entropy loss function, the weightage for vessel classes is 17.63 and for nonvessel it is 1.07. The class weightage is equal to the reciprocal of class occurrence frequency. Class occurrence frequency is the number of pixels associated with the specific class among the total number of pixels.

Databases used for the experimentation are DRIVE [13] and STARE [14]. The images of the DRIVE database were captured using a Canon CR5 nonmydriatic three charge-coupled device camera with a 45 degrees field of view and for STARE, TopCon TRV-50 fundus camera at 35 degrees field of view. DRIVE database is divided equally into training and test set with 20 images each. STARE database is available with twenty images and two independent manual segmentations as ground truth. Instead of using RGB images as input, only resized enhanced green channel is applied as input. Green channel holds more vital information about the vascular region and hence to enhance the not clearly visible vessels, Contrast limited adaptive histogram equalization is applied on the green channel. The proposed network is also attempted through RGB images as input. The vessel extracted is not proper due to unequal illumination and low contrast

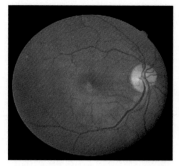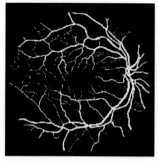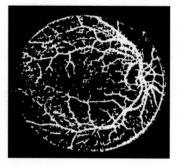

FIGURE 9.4 Vessel extraction with RGB input image.

difference between vessel and background. The vessel extracted by the proposed network for the RGB input is shown in Fig. 9.4.

The images are trained on Matlab 2018 software platform using Intel i5-3570, 32 GB, 64-bit operating system. The proposed fully convolved neural network trains DRIVE database training images for 3500 iterations. Evaluation of the proposed network is done using the performance measures like sensitivity, specificity, and accuracy.

$$\text{Accuracy} = \frac{\text{TP} + \text{TN}}{\text{TP} + \text{TN} + \text{FP} + \text{FN}} \qquad (9.10)$$

$$\text{Sensitivity} = \frac{\text{TP}}{\text{TP} + \text{FN}} \qquad (9.11)$$

$$\text{Specificity} = \frac{\text{TN}}{\text{TN} + \text{FP}} \qquad (9.12)$$

where TP, TN, FP, and FN denote true positive, true negative, false positive, and false negative, respectively. Table 9.3 lists the performance measure for the proposed network for both databases. The proposed fully convolved neural network attained a very high average sensitivity of about 82.37% and 77.67%, respectively, for the DRIVE database and STARE database images. The average accuracy is about 92.61% and 93.1%, respectively, for the DRIVE database and STARE database.

The vessel extracted by the proposed fully convolved neural network is given in Fig. 9.5. Even though the sensitivity achieved is high, it is observed that the correctly classified thick vessels are in huge proportion compared to thin vessels. In addition, it is

Table 9.3 Performance measure for the proposed network.

Database	Sensitivity (%)	Specificity (%)	Accuracy (%)
DRIVE	82.37	94.74	92.61
STARE	77.67	94.31	93.1

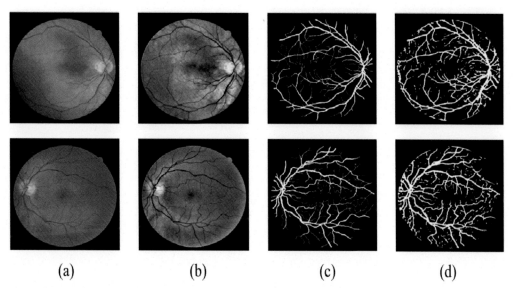

<table>
<tr><td>(a)</td><td>(b)</td><td>(c)</td><td>(d)</td></tr>
</table>

FIGURE 9.5 Blood vessel extraction by proposed network (DRIVE). (A) Color fundus image, (B) green channel image, (C) ground truth, and (D) vessel extracted image.

observed from Fig. 9.5 that the vessel-extracted image is noisier due to the misclassification of background as vessels. The cross-training is achieved by testing abnormal STARE images by the DRIVE training model, and it is shown in Fig. 9.6. It is observed from the STARE images output that the abnormalities like bright and red lesions are well distinguished with the vessel regions. The supervised algorithm for vessel extraction fails in detecting the vessels in the presence of red and bright lesions. Hence, the proposed fully convolved neural network performance is better compared to the above-said method; even the accuracy obtained is less than the other conventional supervised methods.

The proposed fully convolved neural network can be able to distinguish vessels with abnormalities but the limitation is observed that it is inefficient in extracting thin vessels for both normal and abnormal images. More artifacts are observed in the vessel extracted images. The artifacts that are detected as a vessel in the output image can be eliminated by removing the isolated pixels.

9.5.2   Proposed dilated fully convolved neural network

The dilated convolution [15] operation depends on the nonlocal neighbor pixels for its computation. Receptive field is the region of the image that contributes to the specific unit of the network. The significant advantage of dilated convolution is the expansion of the receptive field without any change in number of computation or number of parameters. Usual convolution dealing with the kernel of size m × m is modified into the size of about $m + (m - 1)(r - 1)$, where "r" is the dilated factor. The local neighbor pixels

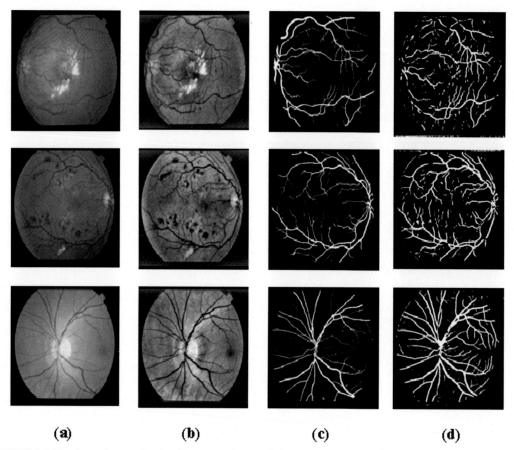

(a)   (b)   (c)   (d)

FIGURE 9.6 Blood vessel extraction by the proposed network (STARE). (A) Color fundus image, (B) green channel image, (C) ground truth, and (D) vessel extracted image.

are considered for the computation of conventional convolution or it is the convolution with a dilation factor as 1. The dilated convolution depends on the dilation factor where the nonlocal neighbor pixels at the specific interval are utilized. The comparison between conventional convolution and dilated convolution is shown in Fig. 9.7.

The vessel extraction can be improved by expanding the receptive field on the image. Expanding the receptive field improves the vessel extraction by considering the vessel region along with the background. The proposed fully convolved neural network in Fig. 9.3 is further enhanced by the use of only the dilated convolutional layers. Table 9.4 shows the performance measure of dilated fully convolved neural network. The dilated convolution is experimented with various dilation factors, and among these, the dilation factor of "2" is better for vessel extraction.

The average accuracy and specificity of DRIVE database are improved with a trivial reduction in sensitivity. As the STARE database comprises only abnormal images,

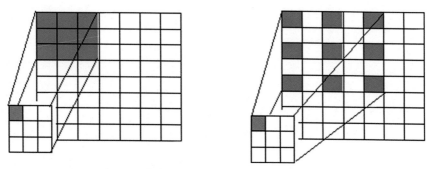

FIGURE 9.7 Dilated convolution.

Table 9.4 Performance measure for dilated fully convolved neural network.

Database	Sensitivity (%)	Specificity (%)	Accuracy (%)
DRIVE	80.73	94.98	93.93
STARE	85.38	90.95	90.61

accuracy is reduced but the network can able to distinguish vessels with abnormalities. It is confirmed through an increase in sensitivity. The vessel extraction result for the DRIVE database and STARE database is given in Figs. 9.8 and 9.9, respectively.

9.6 Artery/vein classification

The fully convolved neural network shown in Fig. 9.9 is proposed for the artery vein classification of the extracted blood vessels. The proposed fully convolved neural network also consists of five stages with two convolutional layers in each stage.

DRIVE AV [13] and IOSTAR [15,16] are the databases used for the experimentation. In the DRIVE AV database, ground truth contains five labels namely artery, vein, background, cross-over, and unlabelled vessels. Labels are shown in red, blue, black, green, and white colors respectively. When the network is trained with all the vascular pixels on the ground truth, the classification result is not observed to be improved. The network trains well when the centreline pixel of ground truth alone is considered for training. In addition, the cross-over, branching points, and unlabelled pixels are ignored for training as well as for testing. Hence, the network is trained with only three labels namely artery, vein, and background. The class imbalance problem is overwhelmed by allocating weights to artery, vein, and background classes. The background is the dominant class, and hence more weightage is given to artery and vein classes. The class weights of background, artery, and vein are 1.0317, 61.8078, and 68.5737, respectively. The number of iterations applied to train the network is about 3500.

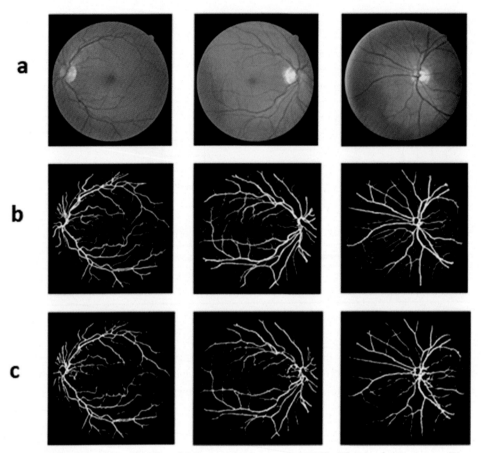

FIGURE 9.8 Blood vessel extracted by dilated fully convolved network (DRIVE). (A) Color fundus image, (B) ground truth, and (C) vessel extracted image.

Table 9.5 shows the accuracy achieved for individual classes for fully convolved neural network-based artery/vein classification. The classified output has only three classes, and only the centerline pixels of the vasculature are labeled. The entire vessel regions with class labels are obtained by inverse distance transform. The isolated pixels are removed to increase the precision of the classification.

The sample outputs shown in Fig. 9.10 are the randomly selected images from DRIVE AV and IOSTAR databases. The alternative nature of the artery–vein vessels is observed to be learned by the network. The accuracy obtained without considering background class accuracy is only 81.89% for the DRIVE AV database. The vessel extracted by the parallel fully convolved neural network [17] is utilized for the postprocessing to intensify the classification performance. The overall classification accuracy with postprocessing is increased to 87.33%. The accuracy for cross-trained result on the IOSTAR database is about 86.18% and 61.78%, with and without background class, respectively.

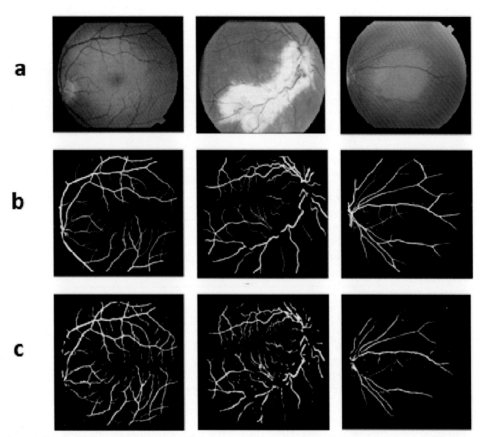

FIGURE 9.9 Blood vessel extracted by dilated fully convolved network (STARE). (A) Color fundus image, (B) ground truth, and (c) vessel extracted image.

Table 9.5   Performance measure of proposed network for artery/vein classification.

Accuracy (%)	DRIVE	IOSTAR
Background	98.2	88.04
Artery	81.66	62.75
Vein	82.11	60.82
Overall (only vessels)	81.89	61.78
Overall (with background)	87.33	86.18

9.6.1   Proposed dilated fully convolved neural network for artery/vein classification

The artery vein classification is further improved by replacing all the convolutional layers of the proposed fully convolved neural network given in Fig. 9.9, by dilated convolutional

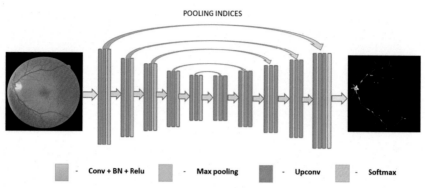

POOLING INDICES

- Conv + BN + Relu   - Max pooling   - Upconv   - Softmax

FIGURE 9.10 Proposed network for artery/vein classification.

Table 9.6 Performance measure for proposed dilated fully convolved neural network for artery/vein classification.

Accuracy (%)	DRIVE	IOSTAR
Background	98.19	89.81
Artery	85.05	67.04
Vein	85.93	59.99
Overall (only vessels)	85.49	63.52
Overall (with background)	89.72	88.12

layers. The training options are maintained as the same as that of the fully convolved neural network for artery/vein classification. The centreline pixels alone are trained for the artery andand vein classes. The class imbalance problem is evaded by assigning class weights to the individual classes. The number of iterations applied to train the network is about 3500. Table 9.6 shows the performance measures of artery/vein classification results for DRIVE AV and IOSTAR databases. The classification accuracy is improved by removing the isolated pixels in the classified output. Fig. 9.11 shows the artery/vein classified image by the atrous fully convolved neural network (Fig. 9.12).

The proposed dilated fully convolved neural network result is better compared to the fully convolved neural network. Even the cross-trained result is low, it is observed that the dilated fully convolved neural network improves the classification performance compared to the fully convolved neural network.

9.6.2   Comparison

The performance of dilated fully convolved neural network is better than fully convolved neural network. The overall accuracy with and without including background class is about 81.89% and 87.33% respectively for the fully convolved neural network. For the dilated fully convolved neural network, the overall accuracy is about 85.49% and 89.72%,

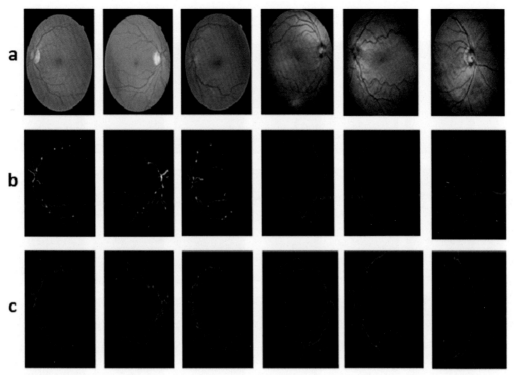

FIGURE 9.11 Artery/vein classified result of the proposed network. (A) Color fundus image, (B) ground truth, and (C) artery vein classified image.

respectively. Due to the isolated misclassified pixels, the output image has some artifacts. This is considerably reduced by the dilated convolution. Furthermore, the alternative nature of artery–vein vessels and the vessel continuity is improved when the convolution layers are replaced by atrous convolution. The comparison of results among the conventional and atrous convolution is listed in Table 9.7.

The proposed fully convolve neural network and dilated fully convolve neural network achieve an average accuracy of about 87.33% and 89.72%, respectively, for the DRIVE AV database. Every pixel in the image is considered for accuracy calculation. The recovery of the entire vascular region from the centreline pixels is done through inverse distance transform. The dilated convolution improves the accuracy of artery, vein, and the overall accuracy calculations compared to a fully convolve neural network for both the databases.

9.7 Summary

In this chapter, a fully convolved deep neural architecture is proposed for retinal blood vessel extraction. The blood vessel extraction is enriched by the replacement of dilated convolution operation instead of conventional convolution in the convolutional layer. The fully convolved neural network and dilated fully convolved neural network are

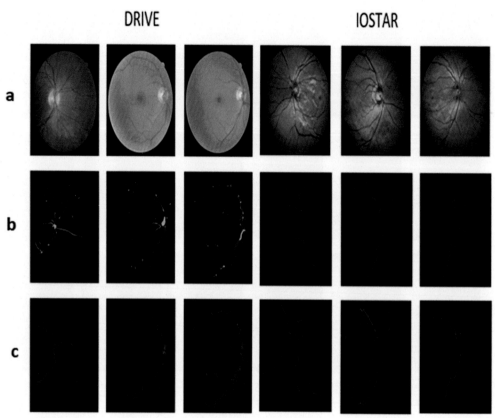

FIGURE 9.12 Artery/vein classified result for proposed dilated fully convolved neural network. (A) Color fundus image, (B) ground truth, and (C) artery vein classified image.

Table 9.7 Comparison table.

Method	Overall	Artery	Vein	Description
	Achieved accuracy (%)			
	DRIVE			
Proposed	87.33 (WB) 81.89 (WOB)	81.66	82.11	Entire pixels in the image for accuracy calculation WB—with background
With dilation	89.72 (WB) 85.49 (WOB)	85.05	85.93	WOB—without background
	IOSTAR			
Proposed	86.18 (WB) 61.78 (WOB)	62.75	60.82	Entire pixels in the image for accuracy calculation WB—with background
With dilation	88.12 (WB) 63.52 (WOB)	67.04	59.99	WOB—without background

proposed for artery/vein classification. The dilated convolutional layer in the proposed architecture improves the artery/vein classification accuracy. For both vessel extraction and artery vein classification, dilation convolution improves the performance. The result obtained by the dilated convolution is better as the field of view of the dilated convolution is enlarged.

References

[1] V. Sathananthavathi, G. Indumathi, BAT algorithm inspired retinal blood vessel segmentation, IET Image Processing 12 (11) (2018) 2075–2083.

[2] V. Sathananthavathi, G. Indumathi, BAT optimization based retinal artery vein classification, Soft Computing 25 (2021) 2821–2835.

[3] Fu, et al., Deepvessel: retinal vessel segmentation via deep learning and conditional random field, in: Proceedings of the International Conference on Medical Image Computing and Computer-Assisted Intervention, 2016, pp. 132–139.

[4] Maninis, et al., Deep Retinal Image Understanding, Medical Image Computing and Computer-Assisted Intervention—MICCAI 2016, vol. 9901, Springer, Cham, 2016.

[5] P. Liskowski, K. Krawiec, Segmenting retinal blood vessels with deep neural networks, IEEE Transactions on Medical Imaging 35 (11) (2015) 2369–2380.

[6] S. Wang, et al., Hierarchical retinal blood vessel segmentation based on feature and ensemble learning, Neurocomputing 149 (2015) 708–717.

[7] Zhou, et al., Improving dense conditional random field for retinal vessel segmentation by discriminative feature learning and thin-vessel enhancement, Computer Methods and Programs in Biomedicine 148 (2017) 13–25.

[8] Fantin, et al., Joint segmentation and classification of retinal arteries/veins from fundus images, Artificial Intelligence in Medicine 94 (2019) 96–109.

[9] R.A. Welikala, et al., Automated arteriole and venule classification using deep learning for retinal images from the UK Biobank cohort, Computers in Biology and Medicine 90 (2017) 23–32.

[10] V. Badrinarayanan, et al., SegNet: a deep convolutional encoder-decoder architecture for image segmentation, IEEE Transactions on Pattern Analysis and Machine Intelligence 39 (2017) 2481–2495.

[11] K. Simonyan, A. Zisserman, Very Deep Convolutional Networks for Large-Scale Image Recognition, ICLR 2015, 2014, pp. 1–14.

[12] M. Ranzato, et al., Unsupervised Learning of Invariant Feature Hierarchies with Applications to Object Recognition, IEEE Conference on Computer Vision and Pattern Recognition, 2007, pp. 1–8. Minneapolis, MN.

[13] J. Staal, et al., Ridge based vessel segmentation in color images of the retina, IEEE Transactions on Medical Imaging 23 (4) (2004) 501–509.

[14] A.D. Hoover, et al., Locating blood vessels in retinal images by piecewise threshold probing of a matched filter response, IEEE Transactions on Medical Imaging 19 (3) (2000) 203–210.

[15] Marios, et al., Semantic segmentation of pathological lung tissue with dilated fully convolutional networks, IEEE Journal of Biomedical and Health Informatics 23 (2) (2019) 714–722.

[16] S. Abbasi Sureshjani, et al., Curvature integration in a 5D kernel for extracting vessel connections in retinal images, IEEE Transactions on Image Processing 27 (2) (2018) 606–621.

[17] S. V, I. G, S.R. A, Parallel architecture of fully convolved neural network for retinal vessel segmentation, Journal of Digital Imaging 33 (2020) 168–180.

10

Grading of diabetic retinopathy using deep learning techniques

Asha Gnana Priya H[1], Anitha J[1], Ebenezer Daniel[2]

[1]*DEPARTMENT OF ELECTRONICS AND COMMUNICATION ENGINEERING, KARUNYA INSTITUTE OF TECHNOLOGY AND SCIENCES, COIMBATORE, TAMIL NADU, INDIA;* [2]*DIAGNOSTIC RADIOLOGY, CITY OF HOPE NATIONAL MEDICAL CENTER, DUARTE, CA, UNITED STATES*

10.1 Introduction

Diabetic retinopathy (DR) is a diabetic complication in which the blood vessels of the retina swell and leak fluids and blood. Diabetes patients who have been sick with the disease for a long time are more likely to have DR. Diabetes wreaks havoc on the retina, heart, nerves, and kidneys. Disease treatment is more effective in the healthcare field, when it is detected at an early stage [1]. Nonproliferative diabetic retinopathy (NPDR) and proliferative diabetic retinopathy (PDR) are the two types of DR, which are detected by the presence of retinal lesions such as microaneurysms (MA), hemorrhages (HM), exudates (EX), and cotton wool spots (CWSs). The NPDR stages are mild, moderate, and severe, with with PDR being the most severe and causing vision loss [2]. MA looks like small red round dots on the retina as a result of vessel wall weakness and is the first sign of DR. There are sharp margins and the size is less than 125 μm, usually between 15 and 60 μm and is difficult to identify due to its small structure [3]. HM appear as larger spots on the retina, which are larger than 125 μm in size and have an irregular margin [4]. EX refers to lipoproteins that leak from detective retinal capillaries. They appear in the outer layer of the retina as tiny white or yellowish-white deposits with sharp borders that are commonly organized in clusters or circinate rings [5]. CWSs appear as white spots on the retina as a result of nerve fiber swelling, and its shape may be oval or round. They are also known as soft exudates that can indicate both advanced NPDR and prePDR [6]. It is difficult to retain the vision in the final stage, so regular retinal screening is required for patients with diabetes to diagnose and treat the early stage of DR to avoid blindness. The importance of DR screening programs, as well as the inability to achieve a stable early diagnosis of DR at a reasonable cost, necessitates the development of a computer-aided diagnosis (CAD) tool. DR can be precisely identified by a trained ophthalmologist, but there are fewer of them. Ophthalmologist screening for DR is prone to error, so accuracy must be improved. Computer-aided disease diagnosis could help clinicians spend their time more efficiently and promote mass screening of the diabetes population in retinal image analysis [7].

CAD technique exploits retinal images for screening of disease. Over the last 160 years, retinal aging has evolved significantly, and it is now an important part of clinical care and therapy for patients with retinal and systemic illnesses. DR, glaucoma, and age-related macular degeneration are detected using fundus photography. In a retina clinic, optical coherence tomography (OCT) and fluorescein angiography are periodically used in the daily management of patients. OCT is also becoming a more useful adjunct in the preoperative and postoperative evaluation of patients undergoing vitreoretinal surgery. The practice of using reflected light to obtain a two-dimensional (2D) representation of the three-dimensional (3D), semitransparent retinal tissues projected onto the imaging plane is recognized as fundus imaging. Thus, fundus imaging refers to any process that yields a two-dimensional image in which the image intensities represent the amount of reflected light. As a result, OCT imaging is not the same as fundus imaging [8]. The retinal fundus image used by the CAD system is less expensive and faster than manual diagnosis. Various image processing, machine learning, and neural network algorithms are used to detect abnormalities in the images. For ophthalmologist visualization, the mathematical representation of features is used to segment and classify retinal lesions. For automatic grading of DR, machine learning and deep learning approaches are used, with multiclass classification implementation proving to be the most difficult task for the researchers. In fundus imaging, deep learning algorithms can be used to highlight important areas and abnormal findings. Deep learning architectures save money while also improving performance [9,10].

The structure of blood vessels in the retina of the eye gives insight into the alterations that occur as a result of retinal illness. Some of the eye features used to evaluate DR and other eye diseases constitute the vascular blood vessels, fovea, and optic disc. Image processing techniques including image enhancement, registration, fusion, segmentation, feature extraction, morphology, and classification help researchers to identify the eye disease. In medical imaging, the image registration technique is used to detect changes and the images captured from assorted angles must be aligned in a single co-ordinate system for successful registration. The progress of combining multiple images into a single image is acknowledged as image fusion, and the process of splitting an image into multiple regions based on color, intensity, and objects is known as segmentation. Image classification is a method of labeling a group of pixels based on grayscale values or other parameters. Image reasoning is a technique used to improve the comprehension of an image's content [11]. Compared to other methods, for detecting retinal lesions matched filter with morphological operation, matched filter algorithm with morphological operation for detecting different types of lesions gives the best differential evaluation. The lesions can be detected after removing blood vessels and optic disc. This method achieves 97.43%, 98.06%, and 98.68% accuracy in detecting lesions such as MA, HM, and EX which is 0.02% and 1.07% more accurate than differential evolution. When compared to the differential algorithm, the matched filter with morphological operation takes 19.82 s to enumerate [12]. Statistical features are extracted and given to the classifier, and to find the severity of HM, the segmentation method is used. Morphological operations are used after feature detection to detect blood vessels and hemorrhages using a segmentation technique. Then, threshold optimization and gray wolf optimization techniques are proposed to achieve maximum accuracy, sensitivity, and specificity performance metrics [13].

An automated system can aid in the rapid detection of diabetic retinopathy, allowing us to easily follow up treatment to avoid further eye damage. EX, HM, and MA lesions are extracted and given to the hybrid like random forest, support vector machine (SVM), logistic regression, k-nearest neighbor, and multilayer perceptron network. The highest accuracy of 82% occurred for the hybrid method along with a precision score of 0.8119, recall score of 0.8116, and F-measure score of 0.8028 [14]. Features extracted show great potential for detecting DRNP and classification. SVM gives 85% accuracy and 95% sensibility for detecting DRNP, which is high compared to other machine learning techniques [15]. Diabetic retinopathy is diagnosed by Bayesian classification, probabilistic neural network (PNN) and SVM by retinal feature extraction. The features like blood vessels and HM are extracted from NPDR images, and EX can be detected using PDR images. The accuracy obtained using 350 fundus images from PNN, bayes classifier, and SVM is 89.6%, 94.4%, and 97.6%. DIARETDB0 dataset images gives accuracy of 87.69%, 90.76%, and 95.38% for PNN, bayes classifier, and SVM [16].

Medical image detection and segmentation with high-performance, low-latency inference deep learning as a core component embedded in and accelerated by graphics processing units (GPU), and convolutional neural network (CNN) intercepts microaneurysms in retinal fundus images. The semantic segmentation algorithm divides the image pixels and determines whether the fundus image is normal or infected. This can help ophthalmologists for classifying different stages of DR. NPDR prediction and productivity are enhanced by the proper training of deep convolutional neural network [17]. Kaggle dataset is used for classification using AlexNet, VggNet, GoogleNet, and ResNet. This can be done along with transfer learning and tuning of hyper parameters to check how these models classify stages of DR, and the accuracy obtained for this is 95.68% [18]. Back propagation neural network, deep neural network, and CNN use Kaggle dataset with 2000 images for classifying five stages of DR. For better results, RGB converted grayscale image is fed with filters, edge detection, binary conversion, median filtering, and morphological processing [19]. Imbalance of image classes and data size increment can be done by the data augmentation techniques. Data augmentation techniques tend to involve translation, flipping, rotation, shear, contrast scaling, and rescaling. This enables the system function effectively [20]. In the training process, the pretrained VGG-16 layers use twofold and cross-fold validation. The proposed CNN, such as pretrained VGG-16, has 16 layers. The CNN of the 512*512 image input was trained, and the CNN weights have been initialized on a lower image resolution. The CNN was subjected to drop-out, and L2 regularization techniques were used to reduce overfitting. CNN uses the Kaggle dataset for training, and Messidor-2 and DR2 datasets are used for testing the process [21].

Data augmentation like cropping and resizing to balance the classes for classification uses the fine tuning of pretrained networks like ResNet-18, VGG-16, GoogLeNet, DenseNet-12l, and SE-BN-Inception. The accuracy obtained in the SE-BN-Inception network is 0.8284 but the lesions are not detected by this work [22]. The model is trained for detecting DR with DIARETDB1 and messidor database. The area under the receiver operating characteristic curve (ROC) of 0.912%−95% and 0.940%−95%. The classification decision is made by the probability map indicating the level of DR. The simple procedure used for calculating a label function for the entire image had the advantage of not requiring any retraining [23]. Binary and multiclass classification improves by using any type of preprocessing as in similar previous works. High values are not obtained in performance metrics but the value range increased for the proposed method. The accuracy obtained for the comparison of WPCNN is 94.23% and 94.46% for AMInceptionV3. Metaplasticity in CNN facilities DR recognition and diagnosis are concluded thus by detecting the discriminatory features improvement [24]. The design and implementation of GPU accelerated deep CNN classify five different stages of DR based on the severity. The quadratic weighted kappa metric single mode CNN accuracy is 0.386, and the ensembling of such case score is 0.3996 [25].

Performance measures for the machine learning and deep learning sensitivity (SE), specificity (SP), accuracy (AC), precision recall, f-measure, and ROC curve help to find the best methods for classification of different stages of DR. These values can be calculated by true positive, true negative, false positive, and false negative. This is known as a confusion matrix and it helps to find the better performance of the network [26]. The capability of the classifier to correctly form the target class is interrupted by the rate of sensitivity, and the specificity rate is illustrated by the capability of classifier separation. The determination of a certain class is evaluated by precision, and the harmonic mean sensitivity is F-score [27].

In the proposed system, ALEXNET, VGG-16, RESENET50, INCEPTION V3, and GOOGLENET convolutional neural networks are compared, and the performance is analyzed. Grading of DR is of two and multiclass based on the retinal fundus images. Section 10.2 explains the materials and methods, Section 10.3 explains all the pretrained convolutional neural networks and the performance analysis, Section 10.4 explains the results and discussion of all the proposed models, and Section 10.5 is conclusion.

10.2 Materials and methods

The proposed system makes use of images from the Indian Diabetic Retinopathy Image Dataset (IDRiD), the first database to represent an Indian population, and it is the only dataset with both typical DR lesions and normal retinal structures annotated at the pixel level. Table 10.1 gives the details of the IDRiD database.

Table 10.1   Details of IDRiD database.

Dataset	IDRiD
Camera	Retinal fundus camera—Kowa VX-10α, 50 degrees field of view
Number of images	516 (training—413, testing—103)
Image resolution	4288*2848
Image format	.jpg (each of 800 kb)
Ground truth	Yes (different lesions, optic disc)
Experimental features	**1.** Distance between lenses with 39 mm was captured from the diabetic patient. **2.** Examined with a noninvasive fundus camera equipped with a xenon flash lamp.
Experimental factors	Mydriasis treated with one drop of tropicamide at a concentration of 0.5%
Data source	Eye clinic, sushrusha hospital building, nanded, (M.S.), India
Description	**1.** Pixel level annotation data (MA, HM, EX, CWS) **2.** Image level disease grading (DR, DME) **3.** Localization of optic disc and fovea
Link	(http://biomedicalimaging.org/2018/challenges/)

IDRiD dataset was made available as part of the "Diabetic Retinopathy: Segmentation and Grading Challenge," which was held in connection with the IEEE International Symposium on Biomedical Imaging (ISBI-2018), in Washington, DC. Fig. 10.1 shows the sample images for different stages of DR. All the experiments are programmed in MATLAB 2021b on an Intel Core i5 @1.4 GHz Quadcore laptop with 8.00 GB RAM.

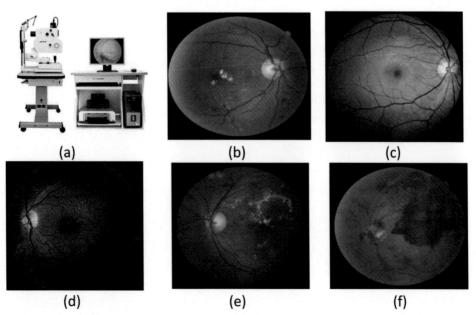

FIGURE 10.1 Fundus camera and different stages of diabetic retinopathy: (A) fundus camera; (B) normal NPDR; (C) mild NPDR; (D) moderate NPDR; (E) severe NPDR; (F) PDR.

10.3 Methodology

CNN training requires more data, and its cost is high compared to transfer learning. The proposed system involves five pretrained models (ALEXNET, VGG-16, RESNET50, INCEPTION V3, GOOGLENET), and the training procedure for all the models is the same. Before training from the retinal fundus dataset, all the pretrained models initialized the weights from the ImageNet database. Convolutional layer, normalization layer, ReLU layer, pooling layer, fully connected layer (FC layer), and SoftMax layer are the layers used in the architecture, and the simple model is shown in Fig. 10.2. Convolutional layer can be any number of layers based on the requirement along with the filter size and number of filters from the argument. Height and width of the filter are represented by the size of the filter. Number of neurons connected to the same input region is represented by the number of filters used. Both the input and output sizes will be the same. To accelerate the network, training normalization layer is placed between the convolution and pooling layer. It is mainly used for speeding the learning process. The nonlinear activation function commonly used is rectified linear unit (ReLU). Output of each convolutional layer improved by this layer by removing zeroes and negative values. The dimensions of the feature map are reduced by the pooling layer that causes faster computation by reducing training parameters. The decision-making layer is the fully connected layer that connects all of the neurons to the neurons in the preceding layer. To classify the images, this layer combines all of the features learned by the preceding layer. This will be the final fully connected layer, and the target will be the same as the output size parameter. This layer activation function is used in the network to normalize it. The classification layer employs classification probabilities determined by the SoftMax function's output, which keeps the sum of positive numbers equal to one. This layer was connected right after the fully connected layer.

All the pretrained network proposed is trained with the stochastic gradient descent momentum at a learning rate of 0.01, and the maximum number of epochs is 300, as specified by the validation data and frequency. After each epoch, the data are shuffled. The network is trained and provides validation accuracy. The unknown images are used in the testing process, and the output is classified as two-class (normal and DR) and multiclass (normal, mild NPDR, moderate NPDR, severe NPDR, and PDR) that defines

FIGURE 10.2 CNN architecture.

DR stages. The last three layers (fully connected layer, SoftMax layer, classification output layer) in the proposed pretrained models can be modified or replaced based on the classification stages by the benefits of transfer learning. The framework of the proposed system is shown in Fig. 10.3.

The performance analysis used in the proposed system is specificity (SP), sensitivity (SE), accuracy (AC), negative predictive value (NPV), positive predictive value (PPV), area under curve (AUC), and receiver operating characteristic (ROC) curve. True positive (TP) and true negative (TN) values are used to calculate all performance metrics. The summary of ROC curve is AUC, which gives the classification stages ability. If the value of AUC is 1, it is considered the classes are correctly classified. If AUC is 0, then the classes are not classified correctly. If the AUC value is between 0.5 and 1, then the classes are partially classified. ROC plot is between the SE and 1-SP of the class detection. The mathematical equations for each performance metrics are given below:

$$AC = \frac{TP + TN}{TP + TN + FP + FN} \tag{10.1}$$

$$SE = \frac{TP}{TP + FN} \tag{10.2}$$

$$SP = \frac{TN}{TN + FP} \tag{10.3}$$

$$PPV = \frac{TP}{TP + FP} \tag{10.5}$$

$$NPV = \frac{TN}{TN + FN} \tag{10.6}$$

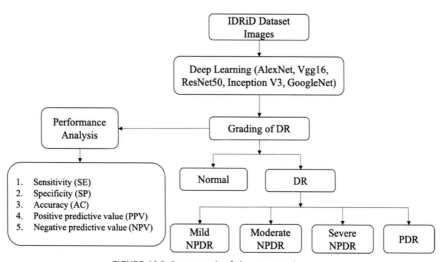

FIGURE 10.3 Framework of the proposed system.

10.3.1 ALEXNET

The pretrained ALEXNET architecture with 25 layers is shown in Fig. 10.4. The input layer of the network is 227*227*3 with zero center normalization. This architecture contains two convolutional layers of 11*11 with stride 2 and zero padding followed by the cross channel normalization with five channels per element, four grouped convolutions of 3*3 with stride 1 and padding 1. Pooling layer used in the network is Maxpooling layer of 3*3 kernels with two strides and zero padding. Finally, three fully connected layers with two 4096 channels and one 1000 channels, which can be changed according to the classification layers, are needed. The validation of two- and five-class of the ALEXNET architecture is shown in Figs. 10.5 and 10.6, which shows the number of epochs and iterations used with the time taken for completing the validation.

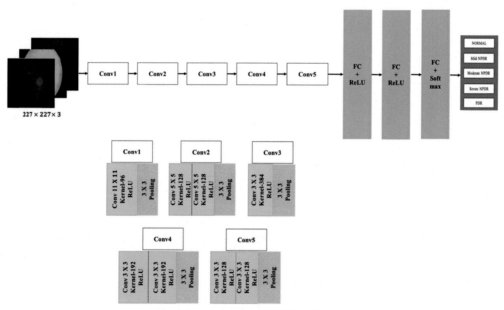

FIGURE 10.4 Pretrained ALEXNET architecture.

Initializing input data normalization.

Epoch	Iteration	Time Elapsed (hh:mm:ss)	Mini-batch Accuracy	Mini-batch Loss	Base Learning Rate
1	1	00:00:00	40.62%	8.4590	1.0000e-04
17	50	00:00:05	78.91%	3.1475	1.0000e-04
34	100	00:00:12	66.41%	5.2675	1.0000e-04
50	150	00:00:19	92.97%	0.7343	1.0000e-04

FIGURE 10.5 Two-class validation of ALENET architecture.

Initializing input data normalization.

Epoch	Iteration	Time Elapsed (hh:mm:ss)	Mini-batch Accuracy	Mini-batch Loss	Base Learning Rate
1	1	00:00:00	27.34%	10.9233	1.0000e−04
17	50	00:00:06	50.78%	7.0365	1.0000e−04
34	100	00:00:12	56.25%	6.5219	1.0000e−04
50	150	00:00:20	93.75%	0.6535	1.0000e−04

FIGURE 10.6 Five-class validation of ALENET architecture.

10.3.2   VGG-16

The pretrained VGG-16 architecture contains 41 layers as shown in Fig. 10.7. The input layer given for this architecture is 224*224*3 with one stride and one padding. Sixteen convolutional layers with kernel 3*3 and stride one are proposed in the network along with five Maxpooling layers with kernel size of 2*2 and one stride. Finally, three fully connected layers with two 4096 channels and one 1000 channels, which can be changed according to the classification layers, are needed. The validation of two- and five-class of the VGG-16 architecture is shown in Figs. 10.8 and 10.9.

10.3.3   RESNET50

The pretrained Resnet50 architecture with 50 layers is shown in Fig. 10.10. The input of resnet50 is 224*224. Resnet is an acronym for residual networks, and the residual block is the central component of the Resnet architecture. Each convolutional block has three convolutional layers with a size of 7*7 followed by a Maxpooling layer with a size of 3*3.

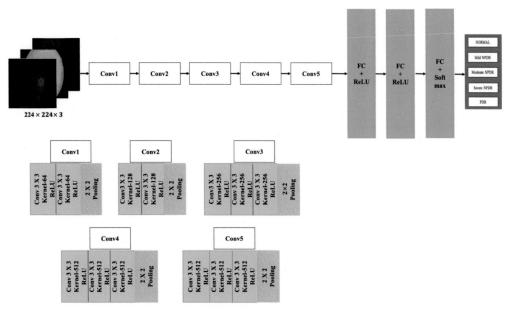

FIGURE 10.7 Pretrained VGG-16 architecture.

```
Initializing input data normalization.
|================================================================================|
| Epoch  | Iteration | Time Elapsed | Mini-batch | Mini-batch | Base Learning |
|        |           | (hh:mm:ss)   | Accuracy   | Loss       | Rate          |
|================================================================================|
|      1 |         1 |   00:00:00   |   65.62%   |  53.0528   |  1.0000e-04   |
|     17 |        50 |   00:00:06   |   71.88%   |   6.9954   |  1.0000e-04   |
|     34 |       100 |   00:00:13   |   80.47%   |   6.6007   |  1.0000e-04   |
|     50 |       150 |   00:00:21   |   89.06%   |   5.5083   |  1.0000e-04   |
|================================================================================|
```

FIGURE 10.8 Two-class validation of VGG-16 architecture.

```
Initializing input data normalization.
|================================================================================|
| Epoch  | Iteration | Time Elapsed | Mini-batch | Mini-batch | Base Learning |
|        |           | (hh:mm:ss)   | Accuracy   | Loss       | Rate          |
|================================================================================|
|      1 |         1 |   00:00:00   |   27.34%   |  10.6391   |  1.0000e-04   |
|     17 |        50 |   00:00:05   |   53.12%   |   6.3649   |  1.0000e-04   |
|     34 |       100 |   00:00:12   |   43.75%   |   7.9378   |  1.0000e-04   |
|     50 |       150 |   00:00:18   |   79.69%   |   2.4022   |  1.0000e-04   |
|================================================================================|
```

FIGURE 10.9 Five-class validation of VGG-16 architecture.

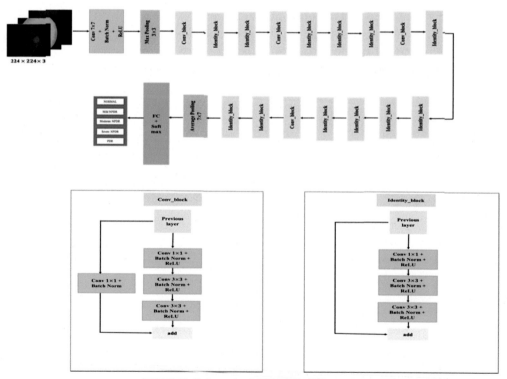

FIGURE 10.10 Pretrained RESNET50 architecture.

```
Initializing input data normalization.
|==============================================================================|
| Epoch  | Iteration | Time Elapsed | Mini-batch | Mini-batch | Base Learning |
|        |           |  (hh:mm:ss)  |  Accuracy  |    Loss    |     Rate      |
|==============================================================================|
|      1 |         1 |   00:00:00   |   28.12%   |   11.2986  |   1.0000e-04  |
|     17 |        50 |   00:00:05   |   51.56%   |    7.0342  |   1.0000e-04  |
|     34 |       100 |   00:00:11   |   73.44%   |    4.0937  |   1.0000e-04  |
|     50 |       150 |   00:00:17   |   86.72%   |    1.7824  |   1.0000e-04  |
|==============================================================================|
```

FIGURE 10.11 Two-class validation of RESNET50 architecture.

```
Initializing input data normalization.
|==============================================================================|
| Epoch  | Iteration | Time Elapsed | Mini-batch | Mini-batch | Base Learning |
|        |           |  (hh:mm:ss)  |  Accuracy  |    Loss    |     Rate      |
|==============================================================================|
|      1 |         1 |   00:00:00   |   29.69%   |   11.0132  |   1.0000e-04  |
|     17 |        50 |   00:00:05   |   50.78%   |    6.6563  |   1.0000e-04  |
|     34 |       100 |   00:00:12   |   55.47%   |    6.5587  |   1.0000e-04  |
|     50 |       150 |   00:00:19   |   78.12%   |    2.3734  |   1.0000e-04  |
|==============================================================================|
```

FIGURE 10.12 Five-class validation of RESNET50 architecture.

Finally, three fully connected layers with two 4096 channels and one 1000 channels, which can be changed according to the classification layers, are needed. The validation of two- and five-class of the RESNET50 architecture is shown in Figs. 10.11 and 10.12.

10.3.4   INCEPTIONV3

The pretrained INCEPTION V3 architecture is a deep network of 48 layers and is shown in Fig. 10.13. The input size of the architecture is 299*299. 1*1 convolutions are added before the larger convolutions to reduce the dimensionality, and the given pooling layer is also as same as the convolutional layer. To improve the architecture's performance, the 5*5 convolutions are divided into two 3*3 layers. N*N convolutions can also be factorized into 1*N and N*1 convolutions. The final layer of the InceptionV3 architecture contains 2048*1000 fully connected layers that can be changed according to the classification layers needed. The validation of two- and five-class of the INCEPTIONV3 architecture is shown in Figs. 10.14 and 10.15.

10.3.5   GOOGLENET

The pretrained GOOGLENET architecture is of deep 22 layers and is shown in Fig. 10.16. The input size of the architecture is 224*224. Inception block in the network is used three times, which includes a convolutional layer of 1*1, 2*2, and 3*3 sizes. Two maximum pooling of 3*3 and one global average pool are used. This architecture consumes less power and less usage of memory. The final layer of the GOOGLENET architecture contains 2048*1000 fully connected layers, which can be changed according to the classification layers needed. The validation of two- and five-class of the GOOGLENET architecture is shown in Figs. 10.17 and 10.18.

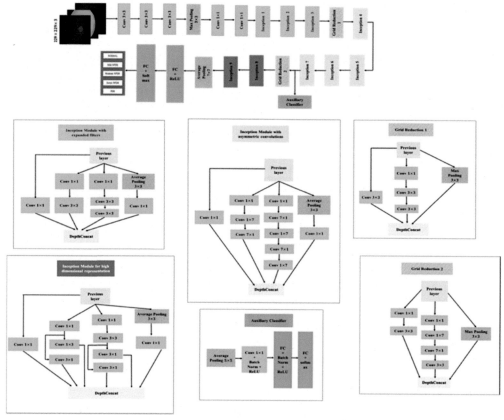

FIGURE 10.13 Pretrained INCEPTIONV3 architecture.

Initializing input data normalization.

Epoch	Iteration	Time Elapsed (hh:mm:ss)	Mini-batch Accuracy	Mini-batch Loss	Base Learning Rate
1	1	00:00:00	67.19%	5.2311	1.0000e-04
17	50	00:00:10	78.91%	2.8780	1.0000e-04
34	100	00:00:22	87.50%	1.7014	1.0000e-04
50	150	00:00:36	93.75%	0.8744	1.0000e-04

FIGURE 10.14 Two-class validation of INCEPTIONV3 architecture.

Initializing input data normalization.

Epoch	Iteration	Time Elapsed (hh:mm:ss)	Mini-batch Accuracy	Mini-batch Loss	Base Learning Rate
1	1	00:00:00	7.81%	14.0166	1.0000e-04
17	50	00:00:13	60.16%	5.8670	1.0000e-04
34	100	00:00:29	57.81%	6.2479	1.0000e-04
50	150	00:00:45	75.78%	3.3682	1.0000e-04

FIGURE 10.15 Five-class validation of INCEPTIONV3 architecture.

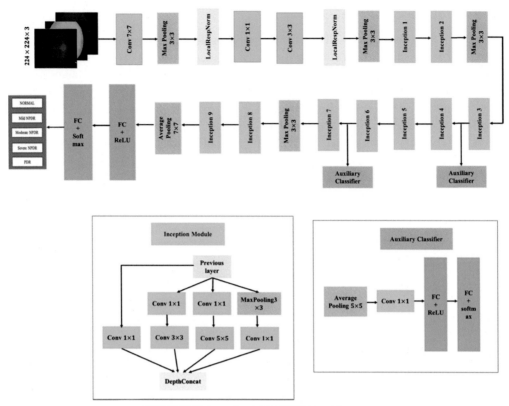

FIGURE 10.16 Pretrained GOOGLENET architecture.

Initializing input data normalization.

Epoch	Iteration	Time Elapsed (hh:mm:ss)	Mini-batch Accuracy	Mini-batch Loss	Base Learning Rate
1	1	00:00:00	34.38%	9.5832	1.0000e-04
17	50	00:00:06	81.25%	2.6662	1.0000e-04
34	100	00:00:12	78.91%	2.9996	1.0000e-04
50	150	00:00:18	82.81%	1.9081	1.0000e-04

FIGURE 10.17 Two-class validation of GOOGLENET architecture.

Initializing input data normalization.

Epoch	Iteration	Time Elapsed (hh:mm:ss)	Mini-batch Accuracy	Mini-batch Loss	Base Learning Rate
1	1	00:00:00	16.41%	11.6966	1.0000e-04
17	50	00:00:06	50.78%	6.6335	1.0000e-04
34	100	00:00:12	61.72%	4.7909	1.0000e-04
50	150	00:00:18	79.69%	1.9021	1.0000e-04

FIGURE 10.18 Five-class validation of GOOGLENET architecture.

10.4 Results and discussion

The pretrained architectures AlexNet, VGG-16, ResNet50, InceptionV3, and GoogleNet use the final classification layers in the fully connected layer. The target set for the classification is also given to this layer and this is the decision-making layer. The parameters used for all the models are almost the same, and the input size and the number of layers may vary based on the network. The proposed parameters are tabulated in Table 10.2, and the performance analysis is given in Table 10.3. The ROC of all the models with two classes is shown in Fig. 10.19, and all models with five classes are shown in Fig. 10.20. From the table and graph, it is clearly known that accuracy for two-class models is higher compared with five-class models. The ROC curve is between the

Table 10.2 Implementation parameters of proposed models.

Parameters	ALEXNET	VGG-16	RESNET50	INCEPTIONV3	GOOGLENET
Size of input	(227,227,3)	(224,224,3)	(224,224,3)	(299,299,3)	(224*224*3)
Weights	ImageNet	ImageNet	ImageNet	ImageNet	ImageNet
Epoch	50	50	50	50	50
Iterations	150	150	150	150	150
Classes	2, 5	2, 5	2, 5	2, 5	2, 5
Optimizer	sdgm	adam	adam	adam	adam
Learning rate	0.0001	0.0001	0.0001	0.0001	0.0001
Dropout	0.5	0.5	0.5	0.5	0.5
Classifier	SoftMax	SoftMax	SoftMax	SoftMax	SoftMax

Table 10.3 Performance analysis of proposed architectures.

Architecture		Performance metrics					
		AC	SE	SP	PPV	NPV	AUC
ALEXNET	2 Class	92%	94%	91%	83.4%	96.9%	0.9708
	5 Class	83.8%	78.4%	86.4%	73.4%	89.3%	0.8672
VGG-16	2 Class	93%	84.3%	97.1%	93.4%	92.8%	0.9458
	5 Class	84%	93.3%	79.6%	68.7%	96.1%	0.9440
RESNET50	2 Class	92%	79.9%	97.8%	94.7%	91%	0.9417
	5 Class	80.9%	97.8%	72.8%	63.3%	98.5%	0.9513
INCEPTION V3	2 Class	87.9%	87.3%	88.2%	78%	93.5%	0.0987
	5 Class	85.7%	59%	98.6%	95.2%	83.3%	0.8364
GOOGLENET	2 Class	92.3%	88.1%	94.3%	88.1%	94.3%	0.9505
	5 Class	87.4%	62.7%	99.3%	97.7%	84.7%	0.8605

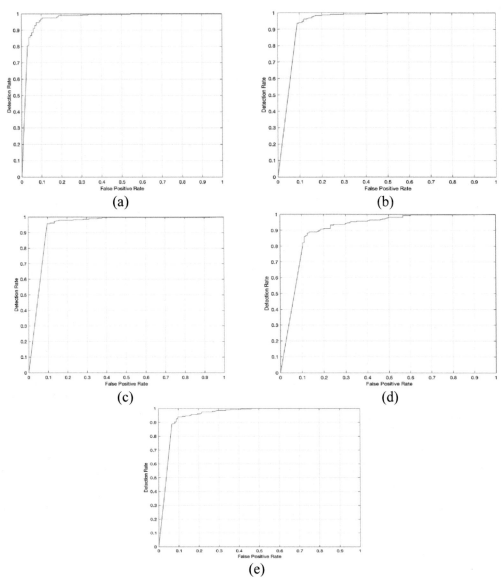

FIGURE 10.19 ROC curves for all the models (two class): (A) ALEXNET; (B) VGG-16; (C) RESNET50; (D) INCEPTIONV3; (E) GOOGLENET.

detection rate and false positive rate. If the values are above the saturation point, the accuracy is higher; thus the performance and classification of the model are considered as the best. From the comparison, nearly all the pretrained models are giving the best performance but the accuracy needs to be increased for clinical purposes.

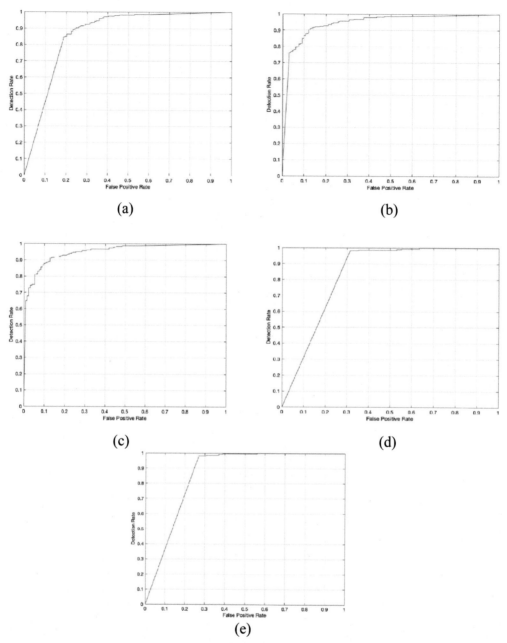

FIGURE 10.20 ROC curves for all the models (five class): (A) ALEXNET; (B) VGG-16; (C) RESNET50; (D) INCEPTIONV3; (E) GOOGLENET.

10.5 Conclusion

DR is the disease of blindness that need proper attention for avoiding vision loss mostly in the diabetes patient. The proposed system implemented five pretrained convolutional neural networks with the IDRiD dataset images as the input. The performance of all the networks is compared, and two-class classification gives more accuracy compared to multiclass classification. Multiclass classification performance has been improved compared to machine learning techniques. Based on the performance analysis, for five-class classification, GOOGLENET gives 87.4% accuracy and for two-class classification, VGG-16 gives 93% accuracy. The proposed method is only on the pretrained networks the model can be modified based on our requirement that may help to increase the performance. Although various methodologies have been established, researchers have yet to generalize a methodology for clinical usage.

References

[1] L.A. Wejdan, et al., Diabetic retinopathy detection through deep learning techniques: a review, Informatics in Medicine Unlocked 20 (2020) 100377.

[2] Q. Imran, et al., Diabetic retinopathy detection and stage classification in eye fundus images using active deep learning, Multimedia Tools and Applications 80 (2021) 11691–11721.

[3] L. Shengchun, et al., Microaneurysms detection in color fundus images using machine learning on directional local contrast, BioMedical Engineering Online 19 (1) (2020) 123.

[4] G.P.H. Asha, et al., Detection and grading of diabetic retinopathy in retinal images using deep intelligent systems: a comprehensive review, Computers, Materials & Continua 66 (3) (2021) 2771–2786.

[5] T. Nipon, et al., Hard exudate detection in retinal fundus images using supervised learning, Neural Computing & Applications 32 (17) (2020) 13079–13096.

[6] S. Syna, Cotton wool spots detection in diabetic retinopathy based on adaptive thresholding and ant colony optimization coupling support vector machine, IEEJ Transactions on Electrical and Electronic Engineering 14 (6) (2019) 884–893.

[7] I.S. Clara, et al., Evaluation of a computer-aided diagnosis system for diabetic retinopathy screening on public data, Investigative Ophthalmology & Visual Science 52 (7) (2011) 4866–4871.

[8] A. Michael, N.K. Christine, Chapter 6-image processing, in: Retina, fifth ed. vol. 1, 2013, pp. 151–176.

[9] B.A. Malik, et al., Identification of diabetic retinopathy through machine learning, Mobile Information Systems 2021 (2021), 1155116, https://doi.org/10.1155/2021/1155116.

[10] L. Vasudevan, et al., Automated detection and diagnosis of diabetic retinopathy: a comprehensive review, Journal of Imaging 7 (9) (2021) 165.

[11] W.K. Muhammad, Diabetic retinopathy detection using image processing: a survey, International Journal of Emerging Technology and Research 1 (1) (2013) 16–20.

[12] R. Alaguselvi, M. Kalpana, Performance analysis of automated lesion detection of diabetic retinopathy using morphological operation, Signal, Image and Video Processing 15 (4) (2021) 797–805.

[13] R. Sasuee, Automatic diagnosis of diabetic retinopathy using morphological operations, International Journal of Sciences: Basic and Applied Research 48 (3) (2019) 213–223.

[14] R. Revathy, Diabetic retinopathy detection using machine learning, International Journal of Engineering Research and Technology 9 (6) (2020) 122–126.

[15] V.C. Enrique, et al., Automated detection of diabetic retinopathy using SVM, in: Proceedings: IEEE XXIV International Conference on Electronics, Electrical Engineering and Computing, INTERCON), Cusco, Peru, 2017, pp. 1–4.

[16] R. Priya, Diagnosis of diabetic retinopathy using machine learning techniques, ICTACT Journal on Soft Computing 3 (4) (2013) 563–575.

[17] Q. Lifeng, et al., Diabetic retinopathy detection using prognosis of microaneurysm and early diagnosis system for non-proliferative diabetic retinopathy based on deep learning algorithms, IEEE Access 8 (2020) 104292–104302.

[18] W. Shaohua, et al., Deep convolutional neural networks for diabetic retinopathy detection by image classification, Computers & Electrical Engineering 72 (2018) 274282.

[19] S. Dutta, et al., Classification of diabetic retinopathy images by using deep learning models, International Journal of Grid and Distributed Computing 11 (1) (2018) 99–106.

[20] P. Chudzik, et al., Microaneurysms detection using fully convolutional neural networks, Computer Methods and Programs in Biomedicine 158 (2018) 185–192.

[21] K. Simonyan, et al., A data-driven approach to referable diabetic retinopathy detection, Artificial Intelligence in Medicine 96 (2019) 93–106.

[22] T. Li, et al., Classification of diabetic and normal fundus images using new deep learning method, Leonardo Electronic Journal of Practices and Technologies 32 (2018) 233–248.

[23] T.Z. Gabriel, et al., Diabetic retinopathy detection using red lesion localization and convolutional neural networks, Computers in Biology and Medicine 116 (2020) 103537.

[24] V.B. Victor, et al., Diabetic retinopathy detection through convolutional neural networks with synaptic metaplasticity, Computer Methods and Programs in Biomedicine 206 (2021) 106094.

[25] D. Darshit, et al., Diabetic retinopathy detection using deep convolutional neural networks, IEEE (2016) 261–266.

[26] U.E. Minhaz, et al., Performance analysis of diabetic retinopathy prediction using machine learning models, in: Proceedings: Sixth International Conference on Inventive Computation Technologies, 2021, pp. 1048–1052. Coimbatore, India.

[27] T. Hassan, et al., Performance analysis of deep neural network based automatic diagnosis of diabetic retinopathy, Sensors 22 (1) (2022) 205.

Segmentation of blood vessels and identification of lesion in fundus image by using fractional derivative in fuzzy domain

V.P. Ananthi, G. Santhiya

DEPARTMENT OF MATHEMATICS, GOBI ARTS & SCIENCE COLLEGE, GOBICHETTIPALAYAM, TAMIL NADU, INDIA

11.1 Introduction

The human eye is usually examined by taking an eye fundus image. But this requires a humpty time for a medical process to analyze an eye using a computer without any automation [1]. An example of a fundus image is shown in Fig. 11.1.

Computer algorithms need to be written in such a way to minimize the time duration while analyzing a human eye digitally [2]. These algorithms also need to be improved for analyzing blood vessels and detecting tumors or diseases without human interruption [3,4]. Usually, medical images are poorly illuminated, so initially, a rich preprocessing method should be applied to enhance the contrast of the region of interest. The main aim of eye fundus image enhancement methods is to spotlight blood vessels and increase the contrast. Various enhancement techniques are available in the literature, namely histogram

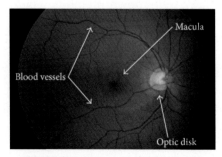

FIGURE 11.1 Human eye fundus image.

Computational Methods and Deep Learning for Ophthalmology. https://doi.org/10.1016/B978-0-323-95415-0.00011-5

equalization, adaptive histogram equalization, contrast limited adaptive histogram equalization, and contrast stretching [5−7].

Blood vessels in fundus images are enhanced under a number of color spaces such as RGB (red, green, and blue), HSV (hue, saturation, value), HIS (hue, saturation, intensity), YIQ (Y luminance-IQ-Chrominance), and CIELAB (LPerceptual lightness A&B for unique colors red, green, blue, and yellow) [1,8]. Finding abnormalities or lesions in the eye is highly important, because if it is not treated in time then it leads to unpredictable damage and even blindness. So, it is necessary to monitor the eye of such a person regularly and provide treatment [9,10].

Detection of lesions in eye fundus images is one of the crucial and time-consuming processes. There are various lesions that occur in the eye fundus image [11]. Ant colony optimization techniques have been used for the segmentation of lesions in Ref. [12]. The deep learning approach has been implemented in Ref. [13] for edema detection. The author in Ref. [14] introduced retinal image assessment using morphological component analysis. The author in Ref. [15] identified exudate lesions with the help of morphological component analysis.

Snake models were used in Ref. [16] for the identification of blood vessels along with these deformable models have been utilized for acquiring higher accuracy. Many algorithms have been available for the removal of optic disk and all other structures of the retina. The clustering algorithm based on the fuzzy k-means and fuzzy c-means has been implemented in Refs. [3,17] for blood vessels and exudate detection. To segment exudates and other nonexudate regions, neural networks have been implemented in Refs. [18,19]. Discriminant functions have been employed in Ref. [20] for identifying lesions and other regions using three features namely color, size, and shape. Results obtained by this method are good only when illumination is good. That is, the results of their method are low while using low-contrast images.

In this chapter, the proposed method is used to detect the blood vessels; in this process, the original RGB image is taken as an input image; after that, the green channel is extracted; then, the mask is generated and is overlapped into the original image Then, this overlapped image is enhanced by fuzzy enhancement technique for enhancing the vessels in the image. Finally, the blood vessel is extracted with the help of fractional derivative and morphological operator. The results obtained by the proposed method are compared with the already well-known methods. The comparison shows that the proposed method extracts vessels in a better way.

In this chapter, a new segmentation process has been initiated to detect tumor. Initially, red and value channels from the input fundus image in RGB and HSV spaces are, respectively, extracted. After that, both red and value channels are fused to get a new image. By using this, a new image optic disk is removed. A fuzzy thresholding technique is applied to detect the lesion in the eye fundus image by choosing three thresholds. Finally, bright lesions are identified as tumor.

In this paper, Section 11.2 deals with preliminary ideas on fuzzy sets. Image preprocessing techniques have been discussed in Section 11.3.

11.2 Preliminary ideas

Let us discuss some basic theories of fuzzy set and image processing operators utilized in this paper.

11.2.1 Fuzzy set

Our aim is to identify a lesion in the eye fundus image. A particular pixel can be classified as a lesion or nonlesion pixel. But there arises a confusion as to whether the pixel belongs to a lesion or nonlesion region. Fuzzy sets have been utilized to handle this confusion.

Let $B = \{b_1, b_2, b_3, \ldots, b_n\}$ be a nonempty finite set. A fuzzy set F_B of B is defined as

$$F_B = \{\langle b, \mu_{F_B}(b)\rangle | b \in B\},$$

where $\mu_{FB}(b) : B \rightarrow [0,1]$ demonstrates the degree of belongingness of $b \in B$, and the degree of nonbelongingness of $b \in B$ can be written in the form of $1 - \mu_{FB}(b)$.

11.2.1.1 Fuzzy image

Let I_B be a digital eye fundus image. A fuzzy image can be written as

$$F_{I_B} = \left\{\langle I_B(i,j), \mu_{F_{I_B}}(i,j)\rangle | 1 \leq i \leq m, 1 \leq j \leq n\right\},$$

where $I_B(i,j)$ is the intensity of the pixel (i,j), and $\mu_{F_{I_B}}(i,j)$ is the membership of the image $I_B(i,j)$.

11.2.2 Image preprocessing

The aim of preprocessing is to improve the image data that reduce unwanted regions and enhance image features. For example, a noisy image can be preprocessed to suppress the noise data without affecting the image feature. There are many preprocessing techniques available in the literature namely contrast adjustment and filtering techniques [21,22]. The RGB color model produces a better result in the segmentation of blood vessels with fractional derivatives [9,23].

11.2.3 Morphological image processing

It is a nonlinear operator that deals with the shape or morphology of an image. It depends upon the ordering of pixel values in a certain neighborhood and does not depend upon their intensities. Hence, it is usually applied in the binary image [24]. The morphological method works with a template called the structuring element.

11.2.3.1 Erosion

The binary image $I(x,y)$ is eroded by a structuring element $J(a,b)$ that generates a new binary image as $I_{erode}(x,y) = I(x,y) \ominus J(a,b)$. In this process, small details are detected and produce a large gap between two different regions. Erosion reduces the size of the object [25,26]. Edges of each different region can be reconstructed by subtracting eroded images from the original image as

$$I_{edge}(x,y) = I(x,y) \ominus I_{erode}(x,y)$$

11.2.3.2 Dilation

A binary image $I(x,y)$ is dilated by a structuring element $J(a,b)$ that generates a new binary image as $I_{dilate}(x,y) = I(x,y) \oplus J(a,b)$. The effect of dilation is the opposite in the sense of erosion. This operation improves a few pixels based on the structuring elements to the interior and exterior boundary of the region [25,26]. This will fill the gaps between various regions.

11.2.3.3 Opening operation

It is a process of erosion of the binary image $I(x,y)$ with a structuring element $J(a,b)$ followed by dilation. $I_{open}(x,y) = (I(x,y) \ominus J(a,b)) \oplus J(a,b)$

This process opens up space between regions joined by thin lines. After that, the region is dilated till it attains the original structure [25,26].

11.2.3.4 Closing

It is a process in which the binary image $I(x,y)$ is initially dilated with a structuring element $J(a,b)$. Then, the result is eroded to produce a new binary image as $I_{close}(x,y) = (I(x,y) \oplus J(a,b)) \ominus J(a,b)$. It helps in filling up the holes in a region [25,26].

11.2.4 Fractional calculus

Fractional derivative is defined by Caputo in Ref. [27] for an $(n-1)$ continuously differentiable function u as

$$D^q u(y) = \frac{1}{\Gamma(n-q)} \int_0^x (x-t)^{n-q-1} \left(\frac{d}{dt}\right) u(t) dt,$$

with zero initial condition. In recent years, authors suggested the use of fractional derivatives for processing images [28–32]. An improved condition for identifying edges is presented in detail in this section. Grunwald Letnikov defined a noninteger order derivative of a continuous function as an extension for a classical integer order derivative. Let for $\forall p \in R$, $I(s) \in [a,s]$ with $a < s, a, s \in R$ be a $q \in Z$ order continuously differentiable function. The noninteger order derivative of order q, where $p > (q)$ is defined as

$$_aD_s^q I(s) = \lim_{h \to 0} \sum_{q=0}^{\left[\frac{s-a}{h}\right]} \frac{(-1)}{h^q} \frac{\Gamma(q+1)}{\Gamma(p+1)\Gamma(q-p+1)} I(s-ph) \qquad (11.1)$$

Eq. (11.1) can be represented as a difference equation for *qth* order derivative as

$$\frac{d^q}{ds^q}I(s) = I(s) + (-q)I(s-1) + \frac{(-q)(-q+1)}{2!}I(s-2)$$

$$+\frac{(-q)(-q+1)(-q+2)}{3!}I(s-3) + \ldots \quad + \frac{\Gamma(n-q)}{\Gamma(n+1)\Gamma(-q)}I(s-n) + \ldots$$

11.3 Proposed method of blood vessel extraction

The following procedure has been utilized to identify blood vessels in eye fundus images. A schematic diagram of the blood vessel extraction is given as a flowchart in Fig. 11.2.

11.3.1 Extraction of green channel in RGB image

During this process, in the RGB eye fundus image, it is clearly seen that the green channel shows the blood vessels better than other red and blue channels [2]. The reason for such representation is that the green channel has neutral saturation between vessels and background, while in the other two channels it is high or under saturated.

11.3.2 Mask generation

The retinal region in the fundus image is identified by generating a mask by applying threshold operation. Usually, a general threshold is applied to the overall image. But in this paper, a separate threshold will be utilized for each channel to generate a mask as follows, for an image I_{RGB},

$$I_{mask}(x,y) = \begin{cases} 1 & \text{if } I_R(x,y) < T_R, I_B(x,y) < T_B, I_G(x,y) < T_G \& I_N(x,y) < T; \\ 0 & \text{otherwise,} \end{cases}$$

where

$$I_N(x,y) = (I_R(x,y) - I_G(x,y)) + (I_R(x,y) - I_B(x,y)) + (I_G(x,y) - I_B(x,y)),$$

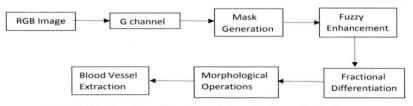

FIGURE 11.2 Flowchart of blood vessel detection in eye fundus image.

T_R, T_B, T_G, and T are the thresholds of red, blue, and green channels of image I_{RGB} and image I_N, respectively. Finally, the generated mask is imposed on the original eye fundus image that is fed into the next process.

11.3.3 Fuzzy image enhancement

In this process, the green channel intensity value of the eye fundus image is mapped into a new gray intensity value using the fuzzy membership function using the following membership function [33].

$$I_{enhance}(x,y) = \begin{cases} 2 * I_G(x,y)^2 \text{if } I_G(x,y) \leq T \\ 1 - 2(1 - I_G(x,y))^2, \text{else if } I_G(x,y) > T \end{cases}$$

where T represents the thresholding of the image I_G.

11.3.4 Edge detection based on fractional differentiation

The enhanced image obtained from the previous section is defuzzied, and the image is differentiated with variable order using the following equation

$$\frac{d^q I_{enhance}(i,j)}{di^q} = I_{enhance}(i,j) + \frac{(-q)I_{enhance}(i-1,j)}{1!} + \frac{(-q)(-q+1)}{2!}I_{enhance}(i-2,j) + \dots$$

$$+ \frac{\Gamma(n-q)}{\Gamma(n+1)\Gamma(-q)}I_{enhance}(i-n,j) + \dots. \frac{d^q I_{enhance}(i,j)}{di^q}$$

$$= I_{enhance}(i,j) + \frac{(-q)I_{enhance}(i,j-1)}{1!} + \frac{(-q)(-q+1)}{2!}I_{enhance}(i,j-2) + \dots$$

$$+ \frac{\Gamma(n-q)}{\Gamma(n+1)\Gamma(-q)}I_{enhance}(i,j-n) + \dots.$$

where $n 6 = 0$ [28,31]. A fractional order mask with order v in eight directions is provided and is represented as Fig. 11.3. The edge image obtained from the fractional differentiation will have overextracted blood vessels. To remove such overextracted vessels, morphological operators have been applied to extract exact blood vessels.

...	0	$\frac{\Gamma(-v+1)}{(v-1)!\Gamma(-v+n)}$	0	...
...	⋮	⋮	⋮	...
...	0	$\frac{(-v)(-v+1)}{2}$	0	...
...	0	$-v$	0	...
...	0	1	0	...

FIGURE 11.3 One directional fractional derivative.

11.3.5   Morphological operator

The edge image obtained by the previous section is consecutively fed into morphological operators such as opening, closing, and erosion. These operators are concertedly applied to extract blood vessels [26]. To reduce the edge of the image area, an erosion operator is applied. To widen the rim of the object, the foreground dilation operation is utilized. After all these processes, blood vessels are identified.

11.4   Proposed method of lesion extraction

Proteins and lipids that busted from blood vessels into the retina through the affected vessel will form exudate [15]. Exudate is a bright lesion that has high contrast as compared to disk that has the same intensity range; a sample image is shown in Fig. 11.4. So, initially, the optic disc area should be removed before identifying the lesion (exudate) from the image. The schematic diagram of the proposed method of lesion extraction in eye fundus image is depicted in Figs. 11.5 and 11.6 shows the 10 eye fundus images from the database of exudates.

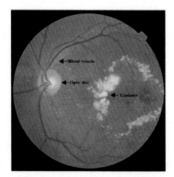

FIGURE 11.4 Exudate in eye fundus image mask.

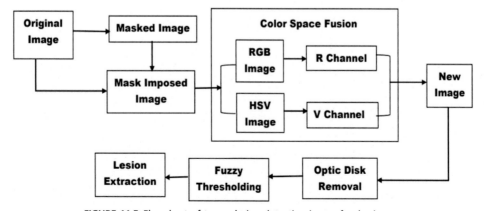

FIGURE 11.5 Flowchart of tumor lesion detection in eye fundus image.

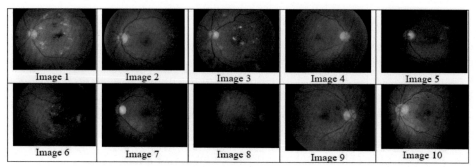

FIGURE 11.6 Original eye fundus images with exudates.

11.4.1 Color space fusion

Initially, the eye fundus image is registered in the computer in RGB space. R component is extracted from RGB space; then RGB eye fundus is then converted into HSV space from which V component is extracted. Finally, R and V components are fused by an averaging operator to get a new image I_F.

11.4.2 Optic disk removal

Morphological operator opening followed by dilation with a sphere structuring element is applied first in this process. Then, the complementary of the output image acquired using morphological operator is superimposed on the fused image I_F to get the optic removed from the original image as shown in Fig. 11.7.

11.4.3 Fuzzy thresholding technique

Lesions like exudate are identified by fuzzy thresholding of the eye image without an optic disc, and its procedure is represented in steps as follows.

Step 1: Initially, RGB eye fundus image I_{RGB} is normalized using fuzzy domain.
Step 2: The RGB image is converted into YIQ color space by using the formula
$Y_{modified} = 1.5 \times Y - I - Q.$

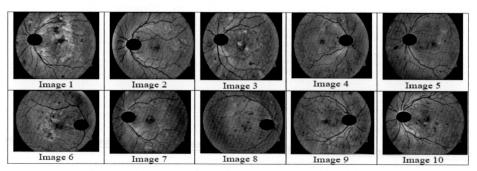

FIGURE 11.7 Optic disk removed eye fundus image in grayscale color space.

Step 3: The I_{RGB} image and $Y_{modified}$ image are fused to produce the new image I_{New}.

Step 4: The new image is fed into a histogram for finding the count of gray levels, and a maximum of three histogram peaks are computed and are noted as T_1, T_2, and T_3.

Step 5: Image in fuzzy domain is thresholded using the following equation:

$$I_{Threshold}(x,y) = \begin{cases} 0, & if I_{New}(x,y) \leq T_1, \\ 0.5 * \left(\dfrac{I_{New}(x,y) - T_1}{T_2 - T_1} \right), & if T_1 < I_{New}(x,y) \leq T_2, \\ 0.5 * \left(\dfrac{T_3 - I_{New}(x,y)}{T_3 - T_2} \right), & if T_2 < I_{New}(x,y) \leq T_3, \\ 1, & else I_{New}(x,y) > T_3. \end{cases}$$

Step 6: Finally, from the thresholded image, $I_{Threshold}$ bright regions are extracted, which are the lesions in the eye fundus image.

11.5 Experimental analysis

Experimentally blood vessel segmentation algorithm is done on a large database STARE and DRIVE [34]. From them, 10 images have been shown in Fig. 11.8. Ground truth and mask images (obtained by proposed method) are shown in Figs. 11.9 and 11.10,

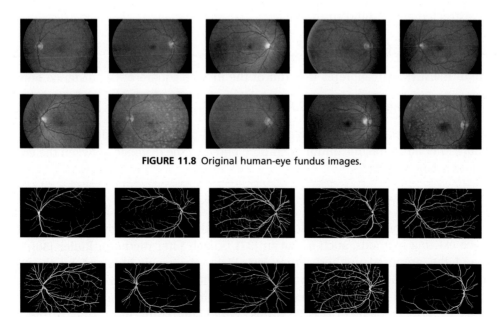

FIGURE 11.8 Original human-eye fundus images.

FIGURE 11.9 Ground truth images.

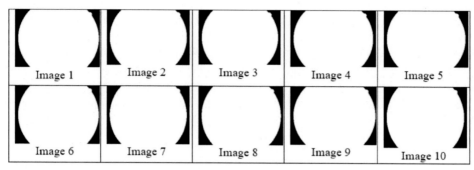

FIGURE 11.10 Mask image.

respectively. To prove the efficiency of the proposed algorithm, the following measures have been utilized.

11.5.1   Precision–recall and receiver operating characteristic graphs

Precision-recall (PR) relates the precision and recall as threshold values changes [35]. Receiver operating characteristic (ROC) plots true positive and false positive rates along varying thresholds [36].

11.5.2   Structural similarity index measure

Images are naturally organized, and it depends upon their structure in them. Structural similarity index measure (SSIM) works by relating luminance (L) and contrast (C) along with structural component (S). $SSIM(I_E, I_R) = L(I_E, I_R)^a . C(I_E, I_R)^b . S(I_E, I_R)^c$, with $0 < a, b, c$. I_E, I_R are extracted and ground truth image.

11.5.3   Figure of merit

Figure of merit (FOM)calibrates the sum of the inverse square of the distance between the original and segmented image. $FOM(I_E, I_R) = \frac{1}{M[N_{I_G}, N_{I_E}]} \sum_{s=1}^{N_{I_E}} \frac{1}{1+\alpha \cdot d_s^2}$, where N_{IG} and N_{IE} are the number of edge points detected from ground truth and edge detected image, α is constant, M denotes the maximum operation, and d_s is the distance between the ground truth and predicted edge.

11.5.4   Result and discussion

Fig. 11.11 shows the segmented blood vessels of the 10 images given in Fig. 11.8 by variance existing methods, such as Sayan [37] (Kirsch filter method), Ridler [38], Tyler [39], and the proposed method. From Fig. 11.11, the result obtained by the Sayan method is oversegmented while compared to ground truth given in Fig. 11.9. Similarly, the results obtained by the Ridler method do not fill the blood vessels, and some other portions are also segmented as edge pixels. While comparing the results of the Tyler method, the result seems to be better than the other two existing methods. But thin

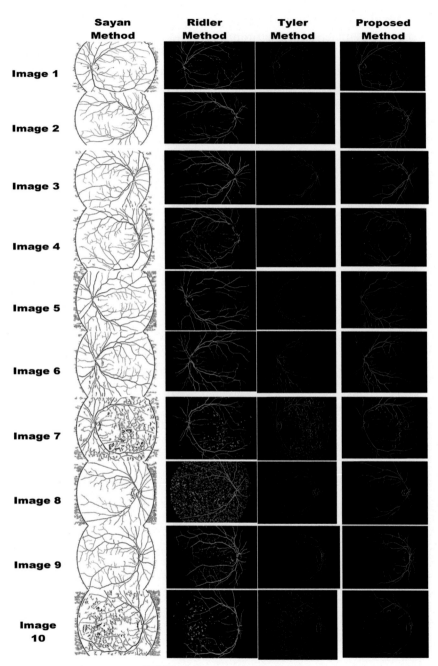

FIGURE 11.11 Segmentation results.

blood vessels are missing. Similarly, in the results of the proposed method, some thin edges are not detected, and it is not oversegmented as the other three methods. So it is visually clear that the proposed method is better in extracting edges while comparing with these three methods.

PR curves for all the 10 images are depicted as pictures from Figs. 11.12–11.21. These all graphs show the efficiency of the proposed vessel extraction method as better when compared to the other three methods. SSIM values of 10 images are drawn as a 3D bar diagram in Fig. 11.22, which also clearly emphasizes the improvement of the proposed method in detecting edges of blood vessels than the other compared methods. Time taken for computation of all the methods has been shown in Fig. 11.23.

Similarly, abnormality segmentation algorithm in eye fundus image is done on large database DIARETDB1, and the ground truth of 10 images given in Fig. 11.6 is shown in Fig. 11.24. These experiments are performed on all the images in the database, but the segmentation results of the images are shown in Fig. 11.25. Experimentally methods are

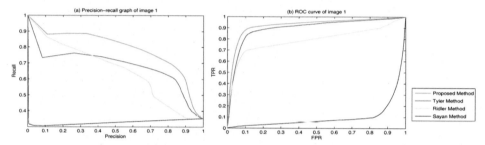

FIGURE 11.12 Precision–recall curves of blood vessel extracted from image 1.

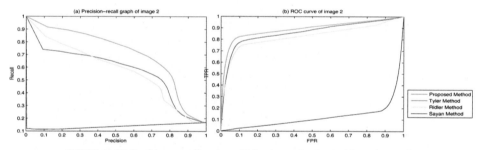

FIGURE 11.13 Precision–recall curves of blood vessel extracted from image 2.

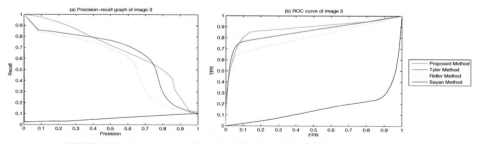

FIGURE 11.14 Precision–recall curves of blood vessel extracted from image 3.

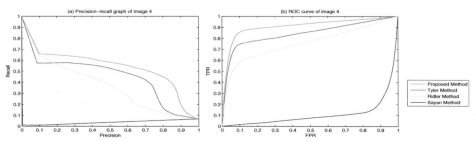

FIGURE 11.15 Precision—recall curves of blood vessel extracted from image 4.

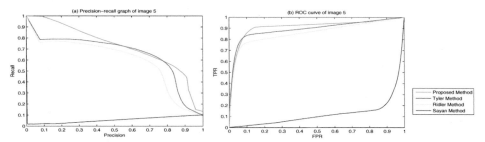

FIGURE 11.16 Precision—recall curves of blood vessel extracted from image 5.

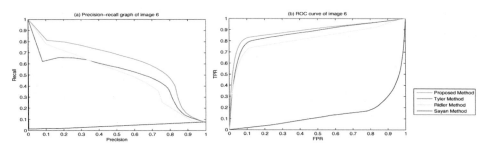

FIGURE 11.17 Precision—recall curves of blood vessel extracted from image 6.

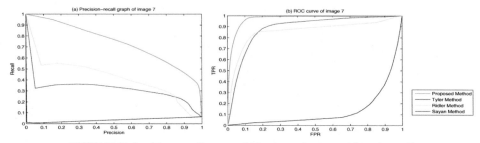

FIGURE 11.18 Precision—recall curves of blood vessel extracted from image 7.

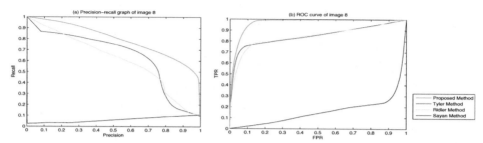

FIGURE 11.19 Precision—recall curves of blood vessel extracted from image 8.

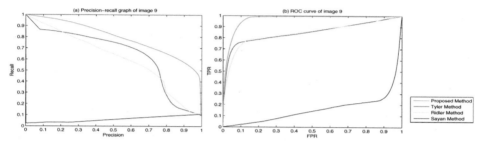

FIGURE 11.20 Precision—recall curves of blood vessel extracted from image 9.

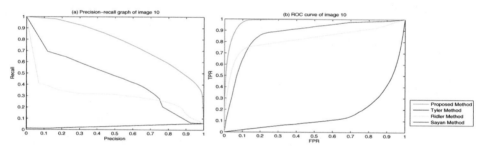

FIGURE 11.21 Precision—recall curves of blood vessel extracted from image 10.

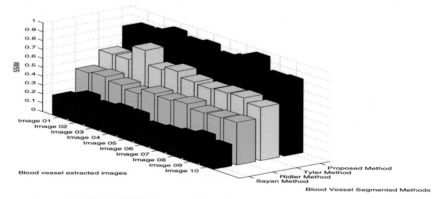

FIGURE 11.22 Values of SSIM for 10 image.

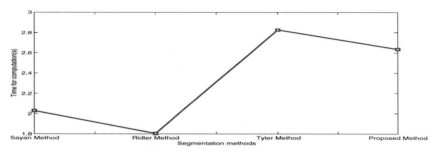

FIGURE 11.23 Computation time in seconds.

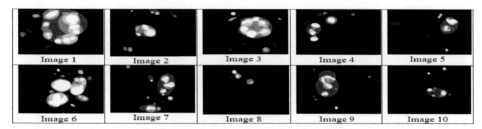

FIGURE 11.24 Ground truth of eye fundus image.

analyzed with the help of MATLAB R2017a. Our proposed method is compared with the MAX-ENT [26], RENENT [40], and SUP-PIX [41] methods.

Fig. 11.25 shows the segmented results of the 10 images depicted in Fig. 11.6. The first and second columns of Fig. 11.25 show the image names and their corresponding original images taken to show segmented output, respectively. The third and fourth columns of Fig. 11.25 show the segmented results obtained using maximum entropy and Renyi entropy methods, respectively. The fifth and sixth columns of Fig. 11.25 show the lesion results acquired by superpixel and the proposed methods. The segmented results of image 1 are shown in the first row acquired using various methods. All the other three-segmented methods show over the segmented region while the proposed method seems to identify properly. For image 2, the results of MAX-ENT and REN-ENT methods are oversegmented, whereas the results of SUP-PIX are undersegmented. One can compare the results obtained from all the methods with the ground truth shown in Fig. 11.24, and among the three existing methods, the proposed method segments the lesion appropriately.

Similarly, for all the rest of the images from 3 to 10, the segmented results obtained using the three existing and the proposed method are depicted in rows in Fig. 11.25. Quantitatively, the proposed method has outperformed other methods since there are more over-segmented regions in the existing methods, but the proposed method is more equivalent to the ground truth. The figures of metric values for all the 10 segmented techniques are evaluated and are shown in Fig. 11.26.

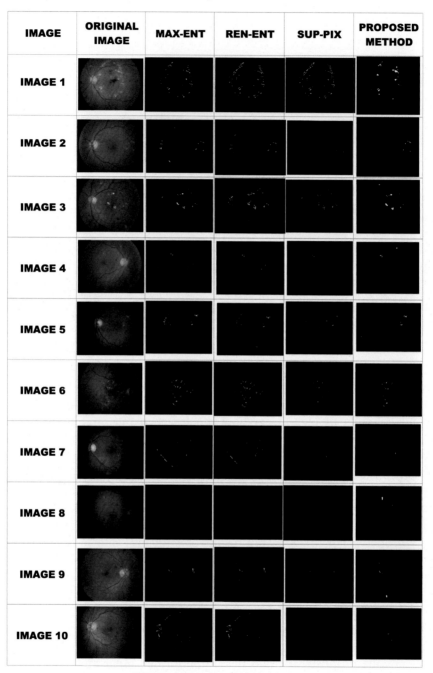

FIGURE 11.25 Segmentation results.

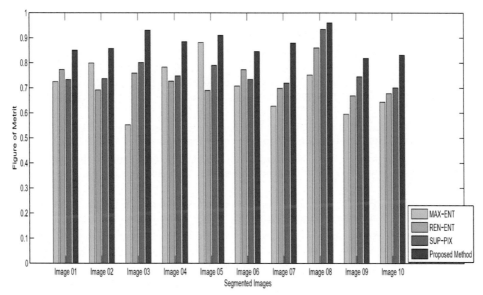

FIGURE 11.26 Figure of merit.

Bar diagram clearly shows the improved results of Renyi entropy over the maximum entropy method, the superpixel method over Renyi entropy method, and the proposed method. Qualitatively and quantitatively the proposed method shows improved performance when compared to the other two compared methods.

11.6 Conclusion

In this chapter, a new edge detection technique has been presented using fractional derivatives in a fuzzy domain. Fuzzy image is initially fed into a fractional order mask for detecting blood vessels. The blood vessels detected from the proposed method seems to be better when compared to the results of other existing technique. Visually and quantitatively, the proposed method is identifying the human eye blood vessels well. Similarly, a new abnormality detection technique was introduced to segment exudates in the eye fundus image. Exudates are extracted by a series of processes from optic disk removal till the detection of the threshold value automatically. The performance of the proposed method extracts exudates more exactly than other comparable methods in terms of both quality and quantity. As a future work, segmentation of tumors will be according to texture or in identifying the type of lesion.

References

[1] M. Patašius, V. Marozas, D. Jegelevičius, A. Lukoševičius, Optimal combinations of color space components for detection of blood vessels in eye fundus images, Elektronika ir Elektrotechnika 91 (3) (2009) 53–56.

[2] M.B. Patwari, R.R. Manza, Y.M. Rajput, N.K. Deshpande, M. Saswade, Extraction of the retinal blood vessels and detection of the bifurcation points, International0 Journal of Computer Application 77 (2) (2013) 29–34.

[3] A. Sopharak, B. Uyyanonvara, S. Barman, Automatic exudate detection from non-dilated diabetic retinopathy retinal images using fuzzy c-means clustering, Sensors 9 (3) (2009) 2148–2161.

[4] R. Naveen, S. Sivakumar, B.M. Shankar, A.K. Priyaa, Diabetic retinopathy detection using image processing, International Journal of Engineering and Advanced Technology 8 (6) (2019) 937–941.

[5] J. Wang, Y.-J. Li, K.-F. Yang, Retinal fundus image enhancement with image decomposition and visual adaptation, Computers in Biology and Medicine 128 (2021) 104–116.

[6] G. Palanisamy, P. Ponnusamy, V.P. Gopi, An improved luminosity and contrast enhancement framework for feature preservation in color fundus images, Signal, Image and Video Processing 13 (4) (2019) 719–726.

[7] K.L. Singh, N.,K. Singh, Histogram equalization techniques for enhancement of low radiance retinal images for early detection of diabetic retinopathy, Engineering Science and Technology, An International Journal 22 (3) (2019) 736–745.

[8] P.J. Navarro, D. Alonso, K. Stathis, Automatic detection of microaneurysms in diabetic retinopathy fundus images using the l* a* b color space, Journal of the Optical Society of America A 33 (1) (2016) 74–83.

[9] Z. Xiao, X. Zhang, L. Geng, F. Zhang, J. Wu, J. Tong, P.O. Ogunbona, C. Shan, Automatic non-proliferative diabetic retinopathy screening system based on color fundus image, BioMedical Engineering Online 16 (1) (2017) 1–19.

[10] F. Qi, G. Li, S. Zheng, Automatic exudate detection in color fundus images, in: International Forum of Digital TV and Wireless Multimedia Communication, Springer, 2016, pp. 155–165.

[11] A. Osareh, M. Mirmehdi, B. Thomas, R. Markham, Classification and localisation of diabetic-related eye disease, in: European Conference on Computer Vision, Springer, 2002, pp. 502–516.

[12] C. Pereira, L. Gonçalves, M. Ferreira, Exudate segmentation in fundus images using an ant colony optimization approach, Information Sciences 296 (2015) 14–24.

[13] J. Sahlsten, J. Jaskari, J. Kivinen, L. Turunen, E. Jaanio, K. Hietala, K. Kaski, Deep learning fundus image analysis for diabetic retinopathy and macular edema grading, Scientific Reports 9 (1) (2019) 1–11.

[14] M. Javidi, A. Harati, H. Pourreza, Retinal image assessment using bilevel adaptive morphological component analysis, Artificial Intelligence in Medicine 99 (2019) 101702.

[15] G. Mahendran, R. Dhanasekaran, N.D. KN, Morphological process based segmentation for the detection of exudates from the retinal images of diabetic patients, in: 2014 IEEE International Conference on Advanced Communications, Control and Computing Technologies, IEEE, 2014, pp. 1466–1470.

[16] L. Espona Pernas, M. Carreira, M. Ortega, M. Penedo, A snake for retinal vessel segmentation, 2007, pp. 178–185.

[17] T. Vandarkuzhali, C. Ravichandran, D. Preethi, Detection of exudates caused by diabetic retinopathy in fundus retinal image using fuzzy k means and neural network, IOSR Journal of Electrical and Electronics Engineering (IOSR-JEEE) (2013) 22–27, e ISSN: 2278–1676.

[18] J.H. Tan, H. Fujita, S. Sivaprasad, S.V. Bhandary, A.K. Rao, K.C. Chua, U.R. Acharya, Automated segmentation of exudates, haemorrhages, microaneurysms using single convolutional neural network, Information Sciences 420 (2017) 66–76.

[19] M. García, C.I. Sánchez, M.I. López, D. Abásolo, R. Hornero, Neural network based detection of hard exudates in retinal images, Computer Methods and Programs in Biomedicine 93 (1) (2009) 9–19.

[20] H. Wang, W. Hsu, K.G. Goh, M.L. Lee, An effective approach to detect lesions in color retinal images, in: Proceedings IEEE Conference on Computer Vision and Pattern Recognition. CVPR 2000 (Cat. No. PR00662), vol. 2, IEEE, 2000, pp. 181–186.

[21] F.F. Wahid, K. Sugandhi, G. Raju, Two stage histogram enhancement schemes to improve visual quality of fundus images, in: International Conference on Advances in Computing and Data Sciences, Springer, 2018, pp. 1–11.

[22] A. Halder, A. Sarkar, S. Ghose, Adaptive histogram equalization and opening operation-based blood vessel extraction, in: Soft Computing in Data Analytics, Springer, 2019, pp. 557–564.

[23] J. Patil, S. Chaudhari, Screening of damage regions in retinopathy using segmentation-color space selection, International Journal Multimedia and Image Processing (IJMIP) 7 (1) (2017) 362–365.

[24] E.R. Dougherty, R.A. Lotufo, Hands-on Morphological Image Processing, vol. 59, SPIE press, 2003.

[25] M.G.F. Eadgahi, H. Pourreza, Localization of hard exudates in retinal fundus image by mathematical morphology operations, in: 2012 2nd International eConference on Computer and Knowledge Engineering (ICCKE), IEEE, 2012, pp. 185–189.

[26] G. Kom, B.W. Tindo, J.M. Pone, A. Tiedeu, et al., Automated exudates detection in retinal fundus image using morphological operator and entropy maximization thresholding, Journal of Biomedical Science and Engineering 12 (03) (2019) 212.

[27] M. Caputo, Linear models of dissipation whose q is almost frequency independent ii, Geophysical Journal International 13 (5) (1967) 529–539.

[28] C. Gao, J. Zhou, Image enhancement based on quaternion fractional directional differentiation, Acta Automatica Sinica 37 (2) (2010) 50–159.

[29] C. Gao, J. Zhou, J. Hu, F. Lang, Edge detection of colour image based on quaternion fractional differential, IET Image Processing 5 (3) (2011) 261–272.

[30] C. Gao, J. Zhou, X. Zheng, F. Lang, Image enhancement based on improved fractional differentiation, Journal of Computational Information Systems 7 (1) (2011) 257–264.

[31] B. Mathieu, P. Melchior, A. Oustaloup, C. Ceyral, Fractional differentiation for edge detection, Signal Processing 83 (11) (2003) 2421–2432.

[32] Y.-F. Pu, J.-L. Zhou, X. Yuan, Fractional differential mask: a fractional differential-based approach for multiscale texture enhancement, IEEE Transactions on Image Processing 19 (2) (2010) 491.

[33] M. Hanmandlu, O.P. Verma, N.K. Kumar, M. Kulkarni, A novel optimal fuzzy system for color image enhancement using bacterial foraging, IEEE Transactions on Instrumentation and Measurement 58 (8) (2009) 2867–2879.

[34] P. Porwal, S. Pachade, R. Kamble, M. Kokare, G. Deshmukh, V. Sahasrabuddhe, F. Meriaudeau, Indian diabetic retinopathy image dataset (idrid): a database for diabetic retinopathy screening research, Data 3 (3) (2018) 25.

[35] M. Sokolova, G. Lapalme, A systematic analysis of performance measures for classification tasks, Information Processing & Management 45 (4) (2009) 427–437.

[36] V.P. Ananthi, Studies on Processing of Images with Uncertainty Using Intutionistic Fuzzy Sets (Thesis), 2017.

[37] S. Chatterjee, A. Suman, R. Gaurav, S. Banerjee, A.K. Singh, B.K. Ghosh, R.K. Mandal, M. Biswas, D. Maji, Retinal blood vessel segmentation using edge detection method, Journal of Physics: Conference Series 1717 (2021) 012008.

[38] T. Ridler, S. Calvard, et al., Picture thresholding using an iterative selection method, IEEE Transactions on Systems, Man, and Cybernetics 8 (8) (1978) 630–632.

[39] T. Coye, A novel retinal blood vessel segmentation algorithm for fundus images, in: MATLAB Central File Exchange, 2015.

[40] D.U.N. Qomariah, H. Tjandrasa, Exudate detection in retinal fundus images using combination of mathematical morphology and renyi entropy thresholding, in: 2017 11th International Conference on Information & Communication Technology and System (ICTS), IEEE, 2017, pp. 31–36.

[41] W. Zhou, C. Wu, Y. Yi, W. Du, Automatic detection of exudates in digital color fundus images using superpixel multi-feature classification, IEEE Access 5 (2017) 17077–17088.

12

U-net autoencoder architectures for retinal blood vessels segmentation

S. Deivalakshmi, R. Adarsh, J. Sudaroli Sandana,
Gadipudi Amarnageswarao

NATIONAL INSTITUTE OF TECHNOLOGY, TIRUCHIRAPPALLI, TAMIL NADU, INDIA

12.1 Introduction

India is the home to one-third of the world's blind population. It has been found out blindness is mainly because of a few causes like age-related macular degeneration, diabetic retinopathy, and glaucoma. If only early diagnosis is possible, then it will be greatly useful to treat these ailments and prevent permanent blindness. Various shape-related parameters such as vessel width, shape, and branching pattern greatly help us in differentiating the affected eye and the normal eye [1]. To assist ophthalmologists in identifying blindness, it is necessary to diagnose the reasons for blindness. Damage to retinal blood vessel leads to blindness can be predicted in advance by successfully segmenting them. To manually segment, characterize and differentiate the retinal blood vessels are tedious and require expertise. As known, manual segmentation is also very time consuming. As there is scarcity of time and expertise, it is very important to automate the segmentation task at hand. So we resort to automatic segmentation. As optical coherence tomography is used to scan the retina, the output will be a digital image on which one have to work on. Hence, image processing techniques were applied on automating the segmentation task in the initial stages [1]. Image processing techniques require the extraction of primary-level features namely morphological traits and vessel width, as well as intermediate-level features like the saliency map. As the level of complexity rises, machine learning and, eventually, deep learning techniques are used to improve efficiency. Thus, later morphological features and saliency maps are extracted and passed through a machine learning model thereby enhancing the segmentation results [2]. Recently, deep learning-based approaches such as convolutional neural network (CNN)-based segmentation have proven their place in medical segmentation because of few advantages like automatic feature extraction and requirement of minimal domain knowledge. Various segmentation-based CNNs such as ALEXNET, GOOGLENET, and VGGNET are examples of CNNs involved in medical image segmentation. Automatic segmentation has been achieved utilizing a variety of techniques, such as image processing, machine learning, and deep learning, as well as computerized medical image segmentation [1] employing image processing techniques. The basic concept is to use retinal images for training the network model and classify patches of images into nerves. Further, Ronneberger et al. [3] introduced the U-net in the field of medical image segmentation. Since then, U-net has become a global pioneer and the most well-known network in medical image segmentation. From then U-net is considered to be one of the benchmarks in medical image segmentation. Residual nets [3] were then brought to medical image analysis to aid in the enhancement of learning in the next layer in deep networks as compared to the previous layers. In this work, variants of U-net also known as autoencoder have been proposed for retinal blood vessels segmentation. The main contribution of this work is the introduction of concatenated filter-based residual paths in the U-net. The concatenated filter-based residual paths reduce the encoder–decoder disparity. The chapter organization is as follows. Related works are given in Section 3. It contains the published works related to retinal blood vessels segmentation. Section 4 describes the proposed work. It contains the explanation on materials and methods used, proposed architecture. Experimental setup is explained in Section 5. It contains patch extraction, processing techniques, and results and discussion. Finally, conclusion is given in Section 6. It contains the conclusion of what the work aims to achieve.

12.2 Related works

Shuang et al. [4] propose a multitask neural network based on a spatial activation method for segmentation of retinal blood vessels using retinal preprocessing and for distinguishing nerves from arteries and veins. Using U-net as an example, Tim et al. [5] image, thereby increasing the speed of the training process dramatically. Jin et al. [6] propose M2U-net, which reduces the number of parameters that need to be learned in high resolution fundus presented DU-net, a deformable U-net that uses deformable convolution to segment retinal blood vessels, resulting in effective context grasps and object localization. Cheng et al. [7] proposed an approach for lung segmentation including eigenvector space and sparse form composition, as well as feature extraction. However, the outcome of the strategy is highly dependent on the complexity of the feature set, which is a significant drawback. It is computationally difficult to extract a feature set that encompasses all anatomical differences across various subjects utilizing eigenvector space and sparse shape attributes. It is also a difficult effort to make sure the model is not overfitted. Gerard et al. [8] proposes a multiresolution CNN that starts with human lung datasets and then expands to other species' datasets. The proposed method, however, allows you to work only on one image at a time and cannot be utilized to discover patterns in several lung images. The approach proposed by Souza et al. [9] contained two levels of segmentation. Patch-based segmentation is used at first, then the image is reassembled into a whole image and segmented again. This strategy outperforms several other options. However, because of the reconstruction process, it generates lot of false positives and reduces true negatives, impacting the quantitative measurements. Wang et al. [10] presented the central focused CNN, a data-driven model that accounts for spatial changes in 2D and 3D images. They also made extensive use of bounding boxes to lessen the segmentation process's computing overhead. Zhang et al. [11] developed a CNN-based network and presented the results of applying the methodology for effective segmentation. In addition, the network was trained with the goal of segmenting retinal blood vessel, and the results were impressive. CNN based network established by Zhang et al. [11] has put forth the results in using the model for effective segmentation. Further the network has been trained with the objective of segmentation of retinal blood vessel. As long as we have a small dataset, Ronneberger et al.'s U-net [3] model for medical image segmentation outperforms other models when it comes to contracting and expanding paths. In terms of medical image segmentation, U-net outperforms due to its ability to recognize traits of global and context scales, although semantic disparity and scale variation issues must be addressed. Attention U-net proposed by Oktay et al. 's [12] has attention gates [AG] to facilitate the segmentation and improves object localization. But there is a need for post processing techniques to effectively improve the performance metrics which is a bit time consuming. Residual network provided by Md Zahongir alom et al. [13] gave an insight on the way residual paths affect the learning objective of the network as compared to the usual convolutional path. The efficiency of building such a path for semantic segmentation has also been discussed. The author also reorganized the well-known U-net, which resulted in better segmentation results. However, the authors propose a novel encoder-decoder fusion technique as a future task. Reza Azad et al. [14] presented the BCDU-net design, which stands for bidirectional convolutional LSTM U-net with densely connected convolutions. Bidirectional convolutional LSTM has been used in conjunction with U-net to improve accuracy through the inclusion of sequential elements such as LSTM.

12.3 Proposed works

12.3.1 Materials and Methods

12.3.1.1 Dataset

Digital retinal images for vessel extraction (DRIVE) dataset has been chosen for experimentation. Four hundred persons have been randomly screened for diabetic retinopathy in Netherlands, and 40 retinal images have been randomly chosen to be a part of DRIVE dataset. Forty images have been divided into 20 for training and 20 for testing the images. Manually segmented images are available as ground truth for training dataset as well as testing dataset.

Fig. 12.1 denotes three subimages: RGB image, Manual annotation 1, and Manual annotation 2. RGB image denotes the original image while Manual annotation 1 and Manual annotation 2 denote the ground truth image segmented by expert

RGB Image   **Manual Annotation 1**   **Manual Annotation 2**

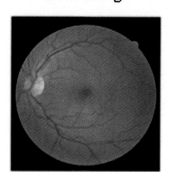 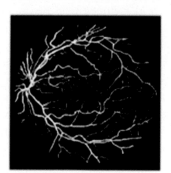 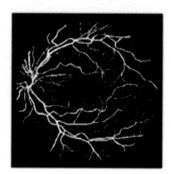

FIGURE 12.1 DRIVE dataset.

ophthalmologists. The DRIVE dataset is created to help comparatively analyze the segmentation of blood vessels in retinal images acting as the benchmark dataset. It does so because it contains various nonlinearities in its dataset in the form of vessel width, varying tortuosity, and height. Every image involved is compressed in JPEG format.

12.3.1.2 Loss function and optimizer

Binary cross entropy [15] is the loss function used in our retinal image segmentation process. It is used since our objective of segmentation is basically a two-class classification problem. The two classes are retinal vessel pixels and

$$H_p(q) = \frac{-1}{N} \sum_{i=1}^{N} y_i \cdot log(p(y_i)) + (1 - y_i) \cdot log(1 - p(y_i)) \qquad (12.1)$$

It is also known as log loss function. Here y_i denotes the class and (y_i) denotes the presence of y class in ith image. N denotes the number of samples to calculate this loss function.

Fig. 12.2 is a representation of variation of log loss or binary cross entropy with respect to variation in the probability of predicted class. It essentially gives us a graphical view of log loss function.

Adam optimizer [16] is used in the segmentation process. It is a combination of adaptive gradient (AdaGrad) and root mean square propagation (RMSprop) optimizer, which eliminates the issue of sparse gradient and helps in reducing noise, respectively. Therefore, Adam optimizer is known to have advantages of both models.

12.3.1.3 Evaluation/Performance Metrics

The performance metrics used in evaluating the performance of the model are global accuracy (AC), area under region of convergence curve (AUC), sensitivity, specificity, and F1-scores. These performance metrics are calculated by comparing the segmented image with the ground truth image.

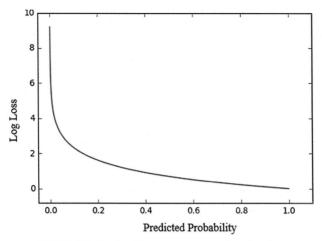

FIGURE 12.2 Log loss/Binary cross entropy loss function.

Global accuracy refers to how accurately each pixel is classified, as shown in Eq. (12.2), that is, how accurately retinal nerve pixels are classified as nerves and nonnerve pixels as background.

$$Accuracy(AC) = \frac{TP + TN}{TP + FP + FN + TN} \qquad (12.2)$$

The capacity of the developed model to correctly recognize nerve pixels and treat them as nerves is referred to as sensitivity, as represented in Eq. (12.3).

$$Sensitivity = \frac{TP}{TP + FN} \qquad (12.3)$$

The ability of the generated model to accurately disregard nonblood vessel pixels and categorize them as background is known as specificity, as represented in Eq. (12.4).

$$Specificity = \frac{TN}{TN + FP} \qquad (12.4)$$

The F1-score is determined by sensitivity and precision given in the Eqs. (12.6) and (12.5), respectively, as shown in Eq. (12.5).

$$F1 - Score = \frac{2 * (Sensitivity * Precision)}{Sensitivity + Precision} \qquad (12.5)$$

where

$$Precision = \frac{TP}{TP + FP} \qquad (12.6)$$

where *TP*—true positive, *TN*—true negative, *FP*—false positive, *FN*—false negative.

12.3.2   Proposed U-net architectures

12.3.2.1   Simple U-net architecture

There are different architectures that have been used to segment retinal blood vessels. It is sort of an evolutionary process. This evolutionary architecture process started with U-net, a benchmark model for medical image segmentation. It is also known as autoencoder that follows encoder–decoder structure. However, parameters like as the number of layers in the encoder and decoder, the number of convolution layers in each layer of the encoder and decoder, the number of layers in the learning module, and hyperparameters are all extensively customizable with U-nets. This ensures with just the U-net generic structure, a lot of models for our application can be generated. One such application is using a regular U-net structure as an initial model for retinal blood vessel segmentation. The generic model shown in Fig. 12.3 consists of four levels in encoder and decoder, which in turn each layer consists of 2 convolution layers. The learning module, which is more popularly known as bottleneck, contains 2 convolution layers.

As mentioned in Fig. 12.3, wherever the green arrow is present, it represents 3×3 convolution with ReLU activation function. Further moving on, the red arrows in the image indicate Max pooling function with a 2×2 window with a stride of 2. So, if basically an image with size A×B with n number of channels then output will be of size $(A/2) \times (B/2)$ with n number of channels. So, using this, both convolutional layer and max pooling layer, the whole working of the U-net can be pointed out. In the encoder, first level receives an image of size 64×64 (which is our input patch size), which then passes through 2 convolutional layers followed by a max pool layer. Thus, output image will be

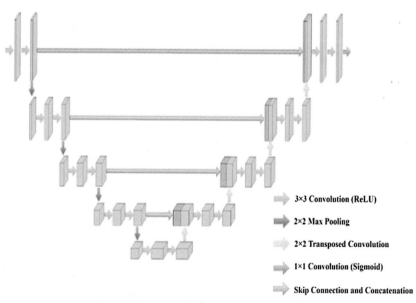

3×3 Convolution (ReLU)

2×2 Max Pooling

2×2 Transposed Convolution

1×1 Convolution (Sigmoid)

Skip Connection and Concatenation

FIGURE 12.3 Simple U-net architecture.

of size 32×32. Similarly, after passing through each level of encoder, the image will become 16 × 16, 8 × 8 and finally 4 × 4 when it reaches the bottleneck structure. On the similar side, in the decoder, we have 4 × 4 getting elevated to 8×8, 16 × 16, 32 × 32 and finally reaching to 64 × 64. The yellow arrow on the decoder side denotes transposed convolution or up convolution. The final blue arrow indicates 1 × 1 convolution with Sigmoid activation function.

12.3.2.2   *Architecture of U-net Variants*

Before getting into a discussion on our architectures, let us list out an important feature that is prevalent in all variants of U-net, which we are able to introduce.

   Medical images have lots of irregular entities like nuclei, boundaries and in this context the nerves. Because of irregularity, the scale of the image varies and hence it affects the segmentation process. Thus, the model proposed should have the capability to overcome the variation of scales problem and should be able to overcome the difficulty caused by the irregularity. U-net tries to do that by skip connections established between encoder and decoder. Even so, the possible semantic gap in learning weights between a U net's encoder–decoder becomes a concern and has an impact on segmentation efficiency. Hence to have correlation between learned weights of the encoder and the decoder and to address the semantic gap problem, residual path components have been introduced as shown in Fig. 12.4A and B. The addition of residual path component into the U-net adapts a regular U-net structure into a model, which reduces the semantic gap between encoder and decoder and also takes care of the irregularities by including the concatenated filters in the skip connection part of the U-net. In the skip connection part, each path consists of variable number of concatenated filters. Each residual path component has one 3 × 3 filter and one 1 × 1 filter, and they are concatenated to feed features to the next component.

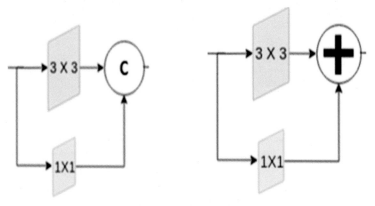

(a) Single residual path component with concatenation   (b) Single residual path component with addition

FIGURE 12.4 Proposed single residual path components.

The residual path component involved in Fig. 12.4A and B contains two convolution filters 3 × 3 and 1× 1 in parallel. The outputs of these two filters are added to produce the final output. This is called a residual path component because it follows a residual path structure, that is, a regular path that consists of 3 × 3 convolution filter and a residual path that consists of 1 × 1 convolution filter. The sections that follow include an overview of the U-net variant architectures that have been proposed.

12.3.2.2.1   ResConU-net achitecture

One such architecture which makes use of both the U-net and residual components above to harness its advantages will be ResConU-net, which is mentioned in Fig. 12.5. As we discussed previously, there is a gap between learned weights of encoder and decoder. This gap keeps on decreasing as we proceed to the bottleneck part of the U-net. Thus, the residual path components are arranged in the fashion of 4, 3, 2, and 1 as they proceed to lower layers as the difference between the learned decoder and juvenile encoder is more at the top layer and reduces when we proceed to lower layers.

12.3.2.2.2   Dense residual path U-net architecture

One more architecture of interest is the dense residual autoencoder shown in Fig. 12.6. It has addition operation instead of concatenation in the residual path component of the residual path. One integral difference between both models lies in the fact that concatenation does not care about the number of channels, and the output from concatenation will be double the number of the channels of the input. But in the case of

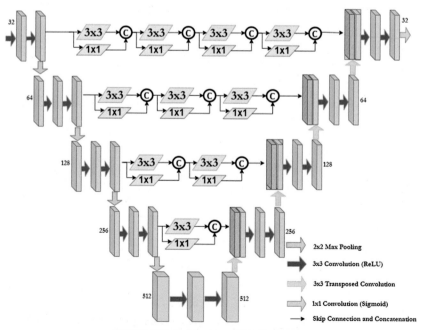

FIGURE 12.5 Architecture of ResConU-net.

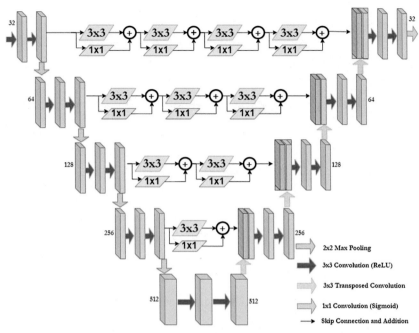

FIGURE 12.6 Architecture of dense residual path U-net.

addition operation, all three dimensions of height, width, and the number of channels of inputs should be equal. Taking this into the notion, the training time is reduced in dense residual autoencoder considering the lesser number of channels. This reduces the number of channels directly and implies a lesser number of training parameters.

12.3.2.2.3   Serial residual U-net architecture

The next variant architecture is cascaded U-net architecture with the residual path in Fig. 12.7. In this network, output of the first network is fed into the output of second network. Even though at first look, it might seem that error can also propagate from first to second network, this network performs better in terms of performance metrics than other networks listed. This is due to the fine tuning of weights carry forward from one network to another. On the flip side, number of training parameters is almost doubled. So, this results in sharp increase of complexity and training time.

12.3.2.2.4   Parallel residual U-net architecture

The next network of interest is shown in Fig. 12.8. This parallel structure resembles the previous cascaded structure, and it involves addition operation to combine the results of the two networks. The output images from the U-nets will be of dimension $64 \times 64 \times 1$. So finally, by adding both of them, the final image will also be of size $64 \times 64 \times 1$. If concatenation is used instead of addition, then the output from concatenation will be of

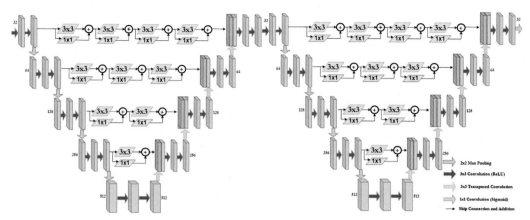

FIGURE 12.7 Serial residual U-net.

size 64 × 64 × 2. So, we will be adding a convolution layer with filter size 1 × 1 and single filter to the output of the concatenation to obtain an image of 64 × 64 × 1.

12.3.2.2.5   Inception block residual U-net architecture

Our final architecture of interest in this section is inception block-based residual autoencoder as shown in Fig. 12.9. The structure is similar to that of ResConU-net discussed earlier with one extra addition of inception block. The inception block is shown in Fig. 12.10. In the convolution layer, filter sizes are fixed at 1 × 1, 3 × 3, 5 × 5, and so on. Filter sizes are varied to capture features at various levels of abstraction. Because an inception block incorporates all the range of filters, it can capture a wide range of spatial fluctuations. In this work, the inception block comprises 3 × 3 and 5 × 5 filters working in parallel. A maximum pooling layer is also constructed in parallel. However, because of the dimensionality disparities, just putting these components together in parallel raises the processing expenses. As a result, each layer receives a 1 × 1 filter to lower the computational costs while enhancing performance. The outputs are then concatenated and fed into the following layer.

12.4   Experiment

12.4.1   Patch extraction and preprocessing

Patch extraction is an integral preprocessing step in retinal blood vessels segmentation. Dividing the images into several subimages called patches is done to localize the segmentation further and to increase the number of training samples fed into the network. The patches of the retinal blood vessel image are shown Fig. 12.11. Fig. 12.11A and B represents the example of the training patch and its corresponding patch in ground truth respectively. Each patch size has been fixed as 64 × 64. From 20 training images, 2,00,000 overlapped subimages have been generated as patches.

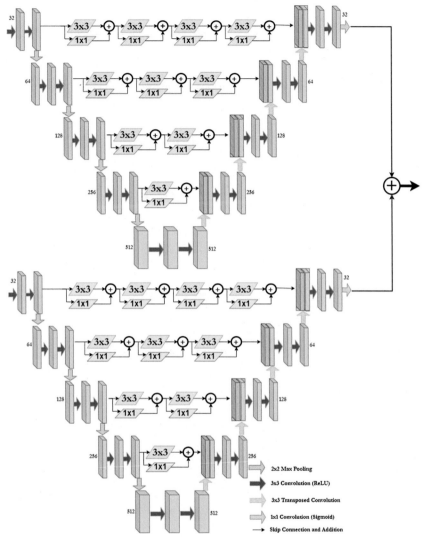

FIGURE 12.8 Parallel residual U-net.

In the test dataset, the images are divided through overlapping using stride height and width of 2 and 2, respectively.

12.4.2   Implementation details

Training of the model is undertaken with the help of Google Collaboratory shortly known as Colab to take advantage of the availability of Graphics Processing Unit (GPU). Python was chosen to take advantage of its packages Keras and Tensorflow. The proposed model is trained on retinal image patches with a size of 64 × 64 and a total of 10,000 subimages

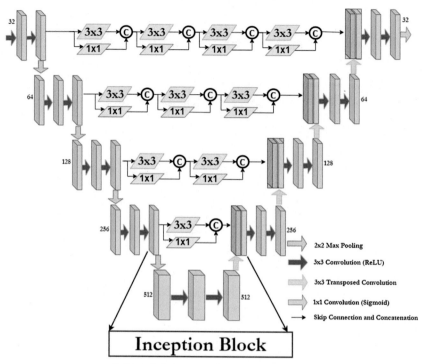

FIGURE 12.9 Inception block residual U-net.

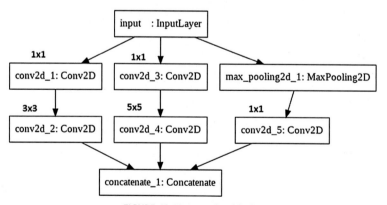

FIGURE 12.10 Inception block.

per image. The number of epochs was randomly chosen with respect to the model. After training for 25 epochs with Adam optimizer and "binary cross entropy" loss function, minimum validation loss is attained at fifth epoch. Then, the trained model is used to evaluate the performance metrics on test images.

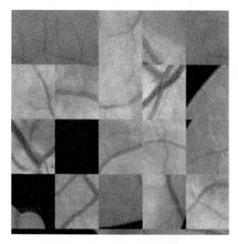
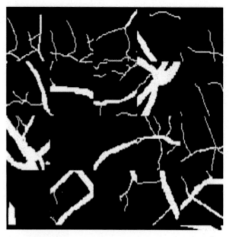

(a) Example of Training Patch   (b) Corresponding patch in ground truth

FIGURE 12.11 Patches of the retinal vessel image.

12.4.3   Experimental results and comparison

Fig. 12.12 shows segmentation results of variants of U-net. The figures follow a grid structure. The x and y axes denote the pixel positions. The top left most pixel represents (0,0) in x and y axes, respectively. Similarly, the bottom right pixel denotes (576,576) which are the last pixels in x and y axes, respectively. Fig. 12.12A represents the segmentation result obtained by ResConU-net. The three images (A), (B), and (C) shown in Fig. 12.12A denotes the original image of the retina, the ground truth image, and the predicted image by our model, respectively. Fig. 12.12B represents the segmentation result obtained by dense residual path U-net. Fig. 12.12C represents the segmentation result obtained by serial residual U-net. Fig. 12.12D represents the segmentation result obtained by parallel residual U-net. Fig. 12.12E represents the segmentation result obtained by inception block-based residual U-net.

From Table 12.1, it can be observed that the performance metrics are better in the proposed networks as compared to the original U-net structure.

Although U-net can overcome the problem of scale variation and the difficulties created by irregularity in segmentation with skip connection, the addition of residual path components improves the model's performance. The proposed residual path component correlates the learned weights and reduces the semantic gap between encoder and decoder. As a result, the model's accuracy in segmentation improves. Also, the model's ability to represent spatial differences of different scales improves with the addition of the inception block. Serial residual U-net, in particular, outperforms the other five U-net variants. The advantage of serial residual U-net is that fine-tuning of weight carries forward from one network to the next, outperforming the previous networks in terms of performance criteria.

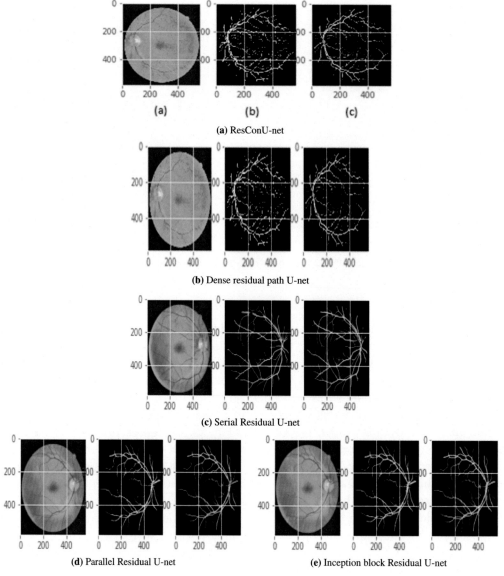

(a) ResConU-net

(b) Dense residual path U-net

(c) Serial Residual U-net

(d) Parallel Residual U-net   (e) Inception block Residual U-net

FIGURE 12.12 Segmentation results of proposed U-net variants.

Overall, the introduction of residual concatenation components in the U-net's skip connection path, which decreases encoder and decoder disparities, as well as the use of inception blocks and cascading the U-net structure, resulted in improved performance.

Table 12.1 Comparison of proposed U-net variants with existing methods.

Methods and year	F1-scores	Sensitivity	Specificity	Accuracy	AUC
Existing methods					
Ronneberger et al. U-net [3]	0.8142	0.7537	0.9820	0.9531	0.9755
Oktay et al. Attention U-net [12]	0.8155	0.7751	0.9816	0.9556	0.9782
Cheng et al. [7]	–	0.7252	0.9798	0.9474	0.9648
Azzopardi et al. [8]	–	0.7655	0.9704	0.9442	0.9614
Liskowski et al. [9]	–	0.7763	0.9768	0.9495	0.9720
Q. li et al. [10]	–	0.7569	0.9816	0.9527	0.9738
Alom et al. Recurrent residual U-net [13]	0.8149	0.7726	0.9820	0.9553	0.9779
Azad et al. BCDU-Net [17]	0.8222	0.8012	0.9784	0.9559	0.9788
Proposed methods					
ResConU-net	0.8199	0.7897	0.9804	0.9561	0.9790
Dense residual path U-net	0.8227	0.7978	0.9793	0.9562	0.9794
Inception block residual U-net	0.8146	0.7713	**0.9822**	0.9560	0.9792
Serial residual U-net	**0.8234**	**0.8156**	**0.9822**	**0.9623**	**0.9832**
Parallel residual U-net	0.8199	0.7878	0.9805	0.9559	0.9790

12.5 Conclusion

For retinal blood vessel image segmentation, U-net variants with autoencoder structures have been proposed. By including concatenation path filters in the U-net, it perceives more information specifically, which results in effective segmentation. By including the inception block, it is able to capture various scales of information. And finally, by cascading the model in series or parallel, we were able to improve some performance metrics; the architectures and the effect of these architectures in terms of results have been summarized and discussed. The experimental results on the DRIVE dataset showed semantic segmentation had a high gain in comparison to other state-of-the-art alternatives. By improving the segmentation results, we will be able to achieve a better diagnosis of eye diseases thus effectively reducing the blindness rate.

References

[1] X. Xu, W. Ding, M.D. Abràmoff, R. Cao, An improved arteriovenous classification method for the early diagnostics of various diseases in retinal image, Computer Methods and Programs in Biomedicine 141 (2017) 3–9.

[2] R. Muzzolinil, Y.H. Yang, R. Pierson, A multiresolution texture segmentation approach with application to diagnostic ultrasound images, in: IEEE Medical Imaging Conference, 2003, pp. 567–578. Santa.

[3] O. Ronneberger, P. Fischer, T. Brox, U-net: convolutional networks for biomedical image segmentation, in: International Conference on Medical Image Computing and Computer-Assisted Intervention, Springer, 2015, pp. 234–241.

[4] W. Ma, S. Yu, K. Ma, J. Wang, X. Ding, Y. Zheng, Multi-task neural networks with spatial activation for retinal vessel segmentation and artery/vein classification, in: International Conference on Medical Image Computing and Computer-Assisted Intervention, Springer, 2019, pp. 769–778.

[5] T. Laibacher, T. Weyde, S. Jalali, M2u-net: effective and efficient retinal vessel segmentation for real-world applications, in: Proceedings of the IEEE/CVF Conference on Computer Vision and Pattern Recognition Workshops, 2019, 0–0.

[6] Q. Jin, Z. Meng, T.D. Pham, Q. Chen, L. Wei, R. Su, Dunet: a deformable network for retinal vessel segmentation, Knowledge-Based Systems 178 (2019) 149–162.

[7] E. Cheng, L. Du, Y. Wu, Y.J. Zhu, V. Megalooikonomou, H. Ling, Discriminative vessel segmentation in retinal images by fusing context-aware hybrid features, Machine Vision and Applications 25 (2014) 1779–1792.

[8] G. Azzopardi, N. Strisciuglio, M. Vento, N. Petkov, Trainable cosfire filters for vessel delineation with application to retinal images, Medical Image Analysis 19 (2015) 46–57.

[9] P. Liskowski, K. Krawiec, Segmenting retinal blood vessels with deep neural networks, IEEE Transactions on Medical Imaging 35 (2016) 2369–2380.

[10] Q. Li, B. Feng, L. Xie, P. Liang, H. Zhang, T. Wang, A cross-modality learning approach for vessel segmentation in retinal images, IEEE Transactions on Medical Imaging 35 (2015) 109–118.

[11] J. Zhang, Y. Cui, W. Jiang, L. Wang, Bloodvesselsegmentationofretinalimagesbasedonneural network, in: International Conference on Image and Graphics, Springer, 2015, pp. 11–17.

[12] O. Oktay, J. Schlemper, L.L. Folgoc, M. Lee, M. Heinrich, K. Misawa, K. Mori, S. McDonagh, N.Y. Hammerla, B. Kainz, et al., Attention U-Net: Learning where to Look for the Pancreas, 2018 arXiv preprint arXiv:1804.03999.

[13] M.Z. Alom, C. Yakopcic, M. Hasan, T.M. Taha, V.K. Asari, Recurrent residual u-net for medical image segmentation, Journal of Medical Imaging 6 (2019) 014006.

[14] L. Liu, Y.Y. Tsui, M. Mandal, Skin lesion segmentation using deep learning with auxiliary task, Journal of Imaging 7 (2021) 67.

[15] K.P. Murphy, Machine Learning: A Probabilistic Perspective, MIT press, 2012.

[16] D.P. Kingma, J. Ba, Adam: A Method for Stochastic Optimization, 2014 arXiv preprint arXiv:1412.6980.

[17] R. Azad, M. Asadi-Aghbolaghi, M. Fathy, S. Escalera, Bi-directional convlstm u-net with densley connected convolutions, in: Proceedings of the IEEE/CVF International Conference on Computer Vision Workshops, 2019, 0–0.

13

Detection and diagnosis of diseases by feature extraction and analysis on fundus images using deep learning techniques

Ajantha Devi Vairamani

AP3 SOLUTIONS, CHENNAI, TAMIL NADU, INDIA

13.1 Introduction

The eyeball is a cystic and ablate spheroid structure. The equator, anterior, and posterior poles of the eyeball are the central points on the greatest convexities of the midplane, anterior, and posterior curvature, respectively. It is made up of anterior and posterior segments, which comprise the front and back parts of the crystalline lens.

The iris, cornea, and anterior and posterior chambers are all filled with aqueous fluid in the anterior section. The vitreous humor-filled area, retina, choroid, and optic disc are all found in the posterior segment [1]. The human eye is the photographic camera's counterpart. It has a crystalline lens, a pupil (variable aperture mechanism), and a retina that looks like film in a camera. The anatomy of the human eyeball is depicted in Fig. 13.1. The rods and cones that are responsible for dark and color views, respectively, are found in the retina, which is the light-sensitive region of the eye [2].

Here are the parts of the human eye that are responsible for vision.

Iris: Iris is a small circular disc that corresponds to the diaphragm of the camera. The pupil, which is located in the center of an iris, controls the amount of light that enters the retina.

Cornea: Cornea is a transparent, avascular structure that resembles a watch glass. The cornea's central third controls three-quarters of the eye's total refractive power.

Sclera: Sclera covers five-sixths of the length of the opaque posterior portion of the eyeball. It helps to keep the shape of the eyeball and protects the eyeball.

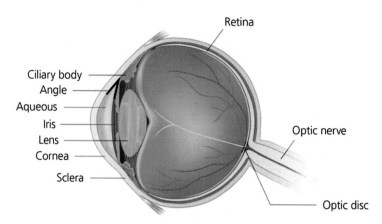

FIGURE 13.1 Anatomy of the human eye ball.

Choroid: The choroid is a blood vessel that supplies blood to the retina and runs between the sclera and the retina. In unusual circumstances, new Blood Vessels (BVs) can form in the choroid area and harm the macula.

Retina: The retina is a thin, sensitive, and transparent membrane that covers the central section of the eyeball. The purplish-red color of the retina is caused by the visual purple of the rods and the underlying vascular choroid. Color differentiation and low-light vision are controlled by rods and cones. The fovea is the macula's central pit, which is responsible for crisp and central vision.

Optical disc: It is a pink-colored circular region with a well-defined outline, measuring 1.5 mm in diameter. Optic cup (OC), also known as physiological cup, is a depression caused by optic disc (OD). BVs have the artery and vein that protrude from the center of this cup.

Optic nerve: The optic nerve starts at the optic disc and ends at the optic chiasm, which is where the two nerves meet. By transporting vision impulses from the retina to the brain, it establishes the first visual system connection.

Crystalline lens: The lens, which sits between the iris and the vitreous humor, is a clear, crystalline, biconvex structure. It is located slightly behind the cornea and is responsible for concentrating images over the retina.

Aqueous humor: Aqueous humor is a clear, watery fluid that fills the anterior and posterior chambers of the body. By removing metabolites from the avascular cornea and lens, it maintains a healthy intraocular pressure (IOP) and optical transparency. It aids in the delivery of oxygen to the eyeball as well as the removal of waste from the lens.

Vitreous humor: Vitreous humor is a colorless, inactive, transparent jelly-like liquid that fills the space between the lens and the retina in the eye. It is the path nutrients travel to reach their destination. It changes from gel to liquid as we age and declines independently of the retina.

13.2 Fundus image analysis

Human retinal images are becoming increasingly useful in the recognition and diagnosis of a wide range of eye infections. Because irregularities in the retina are more obvious than in any other part of the natural eye, retinal photograph diagrams are essential for this application, regardless of the procedure used. A retinal image examination framework is used to diagnose and treat a variety of vision problems. Such diseases include glaucoma, diabetic retinopathy (DR), and age-related macular degeneration, as well as other chronic illnesses that can cause retinal problems, such as cardiovascular and kidney diseases. To avoid vision loss, it is critical to evaluate and treat eye disorders as soon as possible. The burden on ophthalmologists would be considerably reduced.

The significant expense of ebb and flow clinic-based identification strategies (by eye care specialists), as well as a shortage of ophthalmologists, are impediments to those in danger of retinal infections getting the important screening. A potential solution to this problem is to develop a framework for retinal image analysis that makes use of image processing, machine vision, and automated reasoning techniques to identify retinal abnormalities. Various investigations have been led regarding the matter of retinal image examination.

13.3 Eye diseases with retinal manifestation

13.3.1 Glaucoma

Glaucoma is a degenerative eye sickness wherein the optic nerve is gradually destroyed, resulting in vision loss and eventually blindness. It is the second most normal reason for visual impairment in the world. Increased intraocular pressure, or IOP, is the most common cause of damage to the optic nerve's sensitive nerve fibers. Glaucoma can develop in people with a normal IOP in a few circumstances [3]. Glaucoma patients have an IOP greater than 20 mmHg (normal people have an IOP of less than 10 mmHg), which can harm the optic nerve in the posterior region of the eyeball and lead to blindness [4].

The reason for aqueous humor is to nourish the vicinity across the iris and cornea at the same time as additionally exerting stress to maintain the eyeball in form. The fluid is usually created, ensuing in IOP. This inflow is counterbalanced with the aid of using drainage among the iris and cornea, predominantly (80%–90%) through a sponge-like shape referred to as trabecular meshwork, and the final fluids drain independently to hold an IOP. During a watch exam, medical doctors use an ophthalmoscope to peer into the fundus image to test for glaucoma.

13.3.1.1 Classification of glaucoma
Primary open angle glaucoma (POAG) is the maximum not unusual place sort of glaucoma. It is as a result of a sluggish blockage of the drainage canals, ensuing in improved eye stress and optic nerve damage. POAG has an extensive and open perspective among

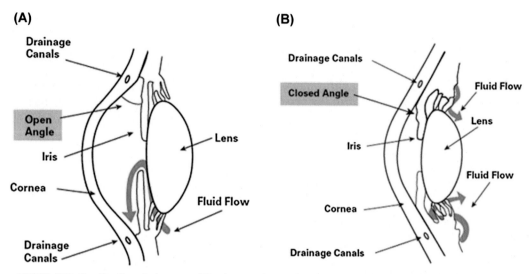

FIGURE 13.2 Classification of glaucoma: (A) primary open-angle glaucoma; (B) primary closed-angle glaucoma.

the iris and the cornea. Primary close angle glaucoma is characterized via way of means of a closed perspective between the iris and the cornea. The iris is found to be impeding aqueous humor drainage via the trabecular meshwork. It is as a result of clogged drainage tubes, which lead to a surprising surge in stress. The extraordinary sorts of glaucoma are depicted in Fig. 13.2.

13.3.2   Diabetic retinopathy

Diabetes is a long-time period situation marked via way of means of excessive blood sugar levels. Diabetes patients, whether or not type I or type II, are at risk of growing DR. In type I and type II diabetes, the frame does now no longer produce sufficient insulin or the cells do now no longer use it properly. If diabetes is detected early, it is able to be dealt with efficaciously and affordably. Diabetics are at risk of imaginative and prescient loss because of DR. A revolutionary discount of blood goes with the drift to lively retinal photoreceptors and a lack of the retina's capillary bed characterizes diabetic retinopathy. DR is characterized as aberrant blood vessel modifications that purpose imaginative and prescient loss with no different symptoms. Diabetic retinopathy is caused by a loss of blood flow and vitamins, including glucose, which causes retinal cells to die and vision to be lost. Diabetes causes renal failure, cardiac disease, and neuropathy by obliterating the capillary bed and tissue blood supply to the kidney, heart, and peripheral nerves [5]. Diabetic retinopathy is a groundbreaking retinal disease characterized by pathological changes in BVs and aberrant fundus lesions such as microaneurysms, hemorrhages, Exudates (EXs), and cotton wool spots (CWSs).

13.3.2.1   Classification of diabetic retinopathy

Proliferative diabetic retinopathy (PDR) and nonproliferative diabetic retinopathy (NPDR) are the two stages of diabetic retinopathy (PDR). The NPDR level, also known as historical DR, is a curable diabetic retinopathy level, whereas the PDR level is the advanced level. Diabetes damages the retinal blood vessels, causing blood, lipid, and protein to seep onto the retina's surface, resulting in NPDR. Due to the leaking, the retina will become wet and bloated, limiting its functionality. Microaneurysms, hemorrhages, EXs, and CWSs are examples of bodily indicators that may be included in the NPDR. The NPDR is divided into three levels based on the number of lesions present: mild, moderate, and severe. The stages of DR are depicted in Fig. 13.3.

13.4 Diagnosis of glaucoma

Glaucoma is a troublesome eye sickness wherein the optic nerve is slowly annihilated by expanded pressure, bringing about fringe vision misfortune.

The Optic Nerve Head Exam is significant for distinguishing glaucoma and checking patients who have been analyzed. Picture signs and imaging modalities like OD and

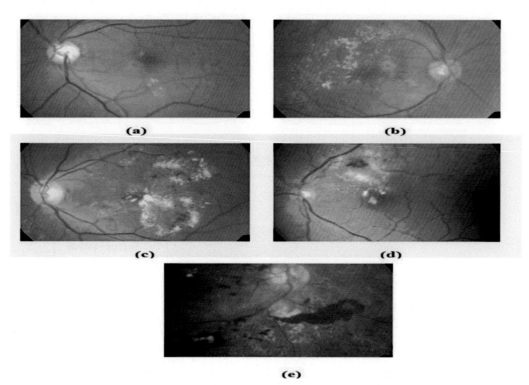

(a) **(b)**

(c) **(d)**

(e)

FIGURE 13.3 The stages of diabetic retinopathy: (A) mild NPDR; (B) moderate NPDR; (C) severe NPDR; (D) very severe NPDR; (E) PDR.

optic cup (OC) are utilized to assemble decision support systems for glaucoma finding. OD is an oval most brilliant locale in the back office of the eye where ganglion nerve strands join to shape the optic nerve head. The OC is a white cup-like area in the focal point of the OD [6]. The Neuro-retinal Rim is the annular zone between the OD and the OC. The primary show of the OD changes when optic nerve filaments are harmed; measuring alludes to the extension of the OC (diminishing of the neuroretinal rim) (Fig. 13.4) [7].

13.4.1   Image acquisition

There are 101 photos in this data set. The patients ranged in age from 40 to 80 years old, and optic disc-centered images with a resolution of 2896 × 1944 pixels were taken. Expert annotated photos were obtained from ophthalmologists with 3, 5, 9, and 20 years of clinical experience.

13.4.2   Localization of optic disc

Preprocessing of retinal images for OD localization is required to account for illumination differences caused by image acquisition in various situations. To locate OD, adaptive histogram equalization is used as a preprocessing step. In addition, the RGB image is grayscale converted. Between the cropped and original images, histogram matching is performed, and the image is then converted to Hue, Saturation and Value (HSV) color space. The correspondence map for each square in the trimmed and unique

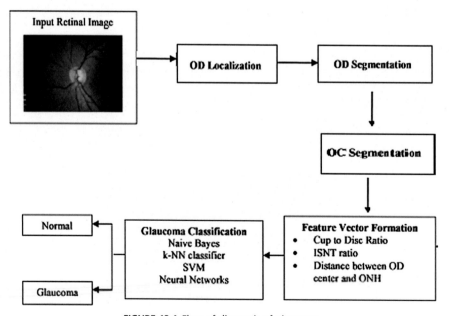

FIGURE 13.4 Flow of diagnosis of glaucoma.

pictures is made utilizing a square distance metric like Euclidean distance. Each square in RM is approximated to a low-layered highlight, which is then used to find the nearest neighbor [8].

The OD is portioned utilizing versatile thresholding. By averaging the base and the most extreme number of lines/segments and working out the column and section focuses, the OD is found. The crossing point of line and section focuses is the OD-center.

13.4.3   Segmentation of optic disc and optic cup

In the OD segmentation, the histogram and centre surround statistics (CSS) are used to classify each pixel as OD or non-OD. Support vector machine (SVM) is used for categorization, which is worth mentioning.

The result esteem is utilized as the evaluation of an incentive for all pixels in the superpixel. Smoothed evaluation values are likewise used to acquire double appraisal values. At last, $+1$, -1, and 0 qualities are appointed to the OD, non-OD, and edge, individually. The subsequent 1s and 0s network is referred to as OD (foreground) and non-OD (background) separately [9]. To portion OC, have utilized similar techniques.

13.4.4   Formation of feature vectors

As demonstrated in the following [10], the feature vector can be created using three prime parameters:
Cup to disc ratio (CDR):

$$CDR = \frac{\text{Area of optic cup}}{\text{Area of optic disc}}$$

Glaucoma is more likely if the CDR is greater than 0.3.
Inferior superior nasal temporal ratio

$$ISNT \ ratio = \frac{\text{Sum of the blood vessels area in inferior and superior region}}{\text{Area of the blood vessels area in nasal and temporal region}}$$

If the inferior superior nasal temporal (ISNT) ratio is low, there is a chance that glaucoma is present [10]. Superior and inferior optic nerve fibers, as well as temporal and nasal optic nerve fibers, are damaged by glaucoma. It causes a decrease in the area of the blood vessel in the superior and inferior neuro-retinal Rim, as well as a shift in the ISNT relationship's order. The recognition of neuro-retinal Rim distances in mediocre, predominant, nasal, and worldly directions is utilized to help the inferior superior nasal temporal rule for early glaucoma analysis [11].

13.4.5   Classifier on glaucoma

The seven classifiers used to classify glaucoma were feed forward back propagation neural networks [10], Naive Bayes [12], SVM [12], k-nearest neighbor (k-NN) [13], distributed time delay neural network [14], radial basis function exact fit (RBFEF) [14], and radial basis function few neurons (RBFFN). Using proper techniques, the retinal images are identified as normal or glaucoma depending on their characteristics. The photos are classified as normal or abnormal using k-NN, SVM, NB, and neural network capabilities. CDR (0.3, 0.3), inferior superior nasal temporal rule (high/low), and DOO (high/low) are statistically significant variables used to classify glaucoma-affected retinal images. The prepared set, example set, and gathering are completely developed utilizing the k-NN classifier. The examples are separated into ordinary and glaucoma cases in light of the prepared set [15]. The input layer contains three nodes, hidden neurons have six, and the output layer has two normal and pathological nodes. RBFEF is utilized for exact information introduction in complex space, while RBFFN is utilized to acquire great Accuracy (ACC) with a couple of neurons inserted [14]. Three elements are applied to the information layer for 26 photographs (13 are typical, 13 are strange), and ordinary and unusual targets are set.

13.5 Diagnosis of diabetic retinopathy

Diabetic retinopathy is a silent illness as the patient may only see the abnormalities in the retina when they have progressed to the point when therapy is complicated and virtually unthinkable. When diabetes first appears and how long it lasts, the extent of retinopathy varies. Until now, the most effective treatment for DR could only be controlled in the early stages of the disease. Early detection through continuous screening is therefore critical. Because it allows for the use of state-of-the-art image processing technologies that automate the detection of anomalies in retinal images, computerized image capturing technology must be used to reduce the cost of such screenings.

DR is a dangerous condition that affects diabetics. It develops as a result of diabetes-related retinal degeneration. It has been proven that early detection and treatment reduce the risk of vision loss and blindness. The fundus camera takes images of the retina, which are then used to diagnose DR. DR screening techniques, as opposed to manual methods of diagnosis, help patients save time, money, and vision. To classify retinal images, various methods have been used [16] used minimum distance discriminant (MDD) classifiers to classify retinal images, using the propagation through radii approach for OD detection [17] used SVM to create a computer-based method for detecting the diabetic retinopathy stage. Image processing techniques were used to extract the features from raw images [18] used the Retinal Grading Algorithm to classify DR intensity automatically based on exudate distribution, count, size, and hemorrhage and microaneurysm distribution [19] classified the data using fractal measurements and clustering algorithms [20] classified retinal images using several different classifiers (Fig. 13.5).

During the categorization process, only the anatomical structures of the optic disc and blood vessels are considered. Mean, variance, entropy, area, diameter, and number of regions extracted from segmented optic discs and from segmented blood vessels are

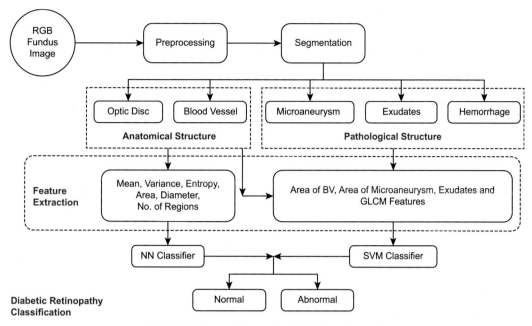

FIGURE 13.5 Flow of diagnosis of diabetic retinopathy.

all input variables for the feed forward neural network classifier. The anatomical structure of blood vessels, hard exudate, and microaneurysm pathological structures, as well as extracted parameters such as vessel pixel intensity, microaneurysm area, exudate area, and textural properties, were analyzed in the second step of the classification process (contrast, homogeneity, correlation, and energy).

13.5.1   Image acquisition

Anatomical components such as the optic disc and blood arteries are segmented from the input retinal images. Using median filtering, the RGB fundus image is preprocessed for noise removal.

13.5.2   Anatomical structure segmentation

Many methods for detecting diseased structures in retinal images require the detection of the basic anatomical structure optic disc as a precondition for the subsequent phases. In retinal image analysis, vessel tracking, measuring distances in retinal images, and documenting changes in the optic disc region due to disease, the location of the optic disc is critical [21]. The OD, which is a bright disc area, is where all major blood vessels and nerves originate [22] used fractal dimension analysis to distinguish the optic disc from other bright regions such as hard exudate because it is the convergence point of blood vessels. Active contour models, Fuzzy C-Means clustering, and artificial neural networks (ANNs) are used to segment the optic disc areas. Images captured by various camera systems in databases were compared to the results of these methods [23]. According to Ref. [24], the normal region growing (RG) method selects seed points by setting an intensity threshold [25] describes an OD localization method that employs a unique circular brightness structure, and OD border detection entails blood vessel inpainting, intensity adjustment, and region growth using the OD center as a seed point. Noise or intensity variations in the RG approach might result in oversegmentation or holes. Furthermore, this approach may not be able to distinguish the shadings of real photographs. To solve these challenges, the modified region growing method utilized in this study takes into account not only the intensity but also the orientation thresholds from the preprocessed images. Seed sites were chosen based on these intensity and orientation threshold traits, and the optic disc was properly divided.

13.5.2.1 Segmentation of optic disc

The photos must be preprocessed before they can be used. The goal of preprocessing is to remove unwanted elements from the image, such as noise, blur, and reflections. To extract only the region of interest (ROI), the retinal images are first trimmed (Fig. 13.6).

After that, the image is converted to grayscale, and the grayscale retinal images are subjected to filtering procedures. The median filter is one of the nonlinear smoothing algorithms employed in this study. The aim behind the median filter is to improve the

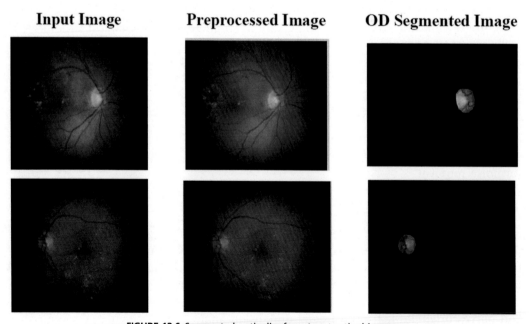

FIGURE 13.6 Segmented optic disc from input retinal images.

outcomes of later processing by replacing the current pixel in the input image with the median of the brightness in its vicinity to maintain edges and decrease noises.

13.5.2.2 Segmentation of blood vessels

The retinal blood vessels, which begin in the center of the optic disc and spread across the entire retinal region, are the second anatomical component visible in the fundus image. Hypertension or blood flow through the vessels in the retina causes injury to these vessels. Fluids and lipids form a ring around injured retinal blood vessels, causing aberrant lesions like exudate and hemorrhages to occur. If these aberrant lesions are identified at an early stage, vision loss in diabetes people can be avoided. As a result, detecting retinal blood vessels is crucial for DR detection (Fig. 13.7).

For the segmentation of blood arteries [26], proposed a multiscale feature extraction and region growth method. In the DRIVE dataset, the authors attained 92.5% accuracy. By doing region-based analysis [27], developed a supervised strategy to extract real vessels. According to Ref. [28], the green channel was extremely responsive to blood vessels [29] proposed an Adaptive Network Fuzzy-Inference System (ANFIS) classifier-based technique for blood vessel segmentation.

13.5.3   Pathological structures segmentation

Microaneurysms, exudate, and hemorrhages are pathological signs to look for when screening diabetic retinopathy [30]. To identify retinal illnesses, it is therefore critical to segregate aberrant characteristics called pathologies in fundus images.

13.5.3.1 Microaneurysms detection

Microaneurysms, among different things, are one of the first indicators of DR [31]. Early detection can assist keep away from irreparable harm to the diabetic eye. The creator evolved a decision-making system, concluding that the location under the chance density feature starting from 0.1 to 0.375 corresponds to the fake reject rate, in which the

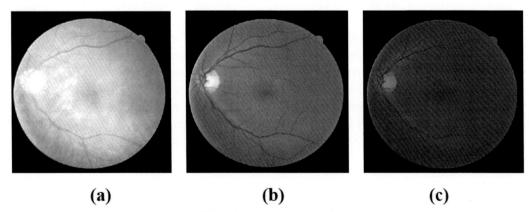

(a)   **(b)**   **(c)**

FIGURE 13.7 Extraction of red, green, and blue channels image: (A) red channel image, (B) green channel image, and (C) blue channel image.

strange man or woman is assessed as regular, and the location starting from 0.375 to 0.4 corresponds to the fake be given rate, in which the regular man or woman is assessed as strange [32] used mathematical morphology to layout a microaneurysm detection system. The set of rules is split into three stages: preprocessing, candidate microaneurysm detection, and postprocessing. According to the authors, the choicest microaneurysm length is among 5 and 16 pixels, with an adaptive histogram equalization threshold of 0.03, and a canny side detection threshold of 0.16.

[33] offered microaneurysm identity primarily based totally on function extraction from an RGB coloration image after morphologically eliminating the optic disc and blood vessels. According to the authors, the recommended technique saves time and strength for ophthalmologists in relation to DR classification.

13.5.3.2 Hemorrhage detection

Hemorrhages are darkish patches that recommend bleeding from blood vessels and are medical signs of retinal illness. Hemorrhage has a black appearance that resembles vessel formations. A retinal hemorrhage is one of the maximum not unusual place signs of nonproliferative DR (NPDR).

The form and quantity of hemorrhages are used to decide the severity of retinal ailment DR [34,35] supplied a powerful approach for detecting crimson lesions called hemorrhages primarily based totally on pixel categorization and mathematical morphology, which has a sensitivity charge of 100% and a specificity charge of 91%. For computerized screening of DR the usage of the SVM classifier, the Splat and Gray Level Co-Occurrences Matrix (GLCM) capabilities are applied [36]. The detection of crimson lesions making use of morphological operations at the inexperienced channel image, which presents the best history for hemorrhage detection, has been proposed by Ref. [37].

13.5.3.3 Exudate detection

Exudate lesions seem withinside the retinal image because of harm to the retinal blood vessels. Ignoring those lesion symptoms and symptoms results in imaginative and prescient loss due to the fact they are hard to hit upon and require analysis at an advance stage. The multiplied stress in blood vessels damages the skinny blood vessels, inflicting them to burst and leak lipids and proteins across the retina. The aberrant lesions withinside the retina are shaped through those lipids. Exudate, cotton wool patches, microaneurysms, and hemorrhages are one-of-a-kind styles of aberrant lesions.

Exudates seem as yellow-white spots of various dimensions and shapes scattered at some point of the retina. Hard and soft exudates are the two kinds of exudates. Hard exudates are intraretinal fatty lesions that might be a key symptom of DR. The identity of exudates is focused, which correlates the DR to save diabetic sufferers from dropping imaginative and prescient sooner. On the opposite hand, detecting exudate for a massive range of pics every 12 months is a totally high-priced and time-ingesting operation that could bring about the human mistake in exudate detection.

13.5.4   Feature extraction from optic disc

The capabilities mean, variance, entropy, and area are retrieved from segmented optic discs of retinal images. In those photos, pixel values are represented as I p, and traits like mean, variance, and entropy are decided for those pixels.
   Mean

$$M_k = \frac{1}{n} \sum_{i=1}^{n} p_i$$

Variance

$$V_k = \left(\frac{1}{n} \sum_{i=1}^{n} (p_i - M_k)^2 \right)$$

Entropy

$$E_i = -\sum_{i} p_i \log_2 p_i$$

   In the previous equations, k denotes the R, G, or B additives of the RGB layout photo, respectively. For every RGB thing of the photos, mean, variance, and entropy are decided independently.
   Area (A) is a unit of dimension that expresses the dimensions of a two-dimensional surface or shape withinside the plane. The area of the photo can be decided on these paintings through counting the full variety of pixels withinside the place. If the segmented place has N pixels, the area (A) is decided as N. Thus, 10 capabilities are retrieved from segmented optic discs of retinal images.

13.5.5   Feature extraction from blood vessels

The functions mean, variance, entropy, area, diameter, and a wide variety of areas are recovered from segmented blood vessel snapshots of retinal imaging. The diameter and wide variety of regions are extracted as characteristics.
   Diameter (*D*)

$$D = \sqrt{\frac{4 \times A}{\pi}}$$

where *A* denotes the area of the image.

Number of regions: Regions can be found in segmented blood vessel retinal images. One of the characteristics of the method is the amount of segmented regions in these segmented images.

As a result, the segmented blood vessels of retinal images yield a total of 12 characteristics. As a result, a total of 22 characteristics are recovered for further classification: 10 features from the optic disc segmented image and 12 features from the blood vessel segmented image.

13.5.6 Classifier on diabetic retinopathy

The functions retrieved from anatomical systems are then used to divide fundus pix into regular and pathological categories. In this study, the feed forward back propagation neural network classifier is hired as one of the everyday class methods. With n I enter nodes, n HU hidden nodes, and n O output nodes, the neural community is a three-layer traditional classifier. When hidden layers are hired, the primary hidden layer is used to accomplice each pair in a single vital unit, and the second one hidden layer is taken into consideration the real hidden layer as soon as the primary hidden layer has categorized the enter data (Fig. 13.8).

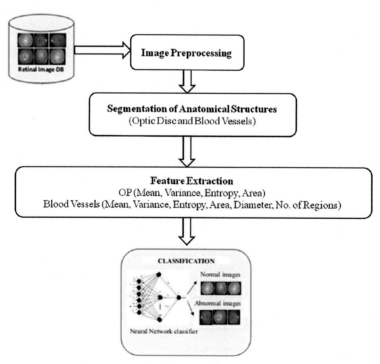

FIGURE 13.8 Classifier on diabetic retinopathy.

13.6 Conclusion

The artificial intelligence device was created to stumble on extreme retinal ailments consisting of diabetic retinopathy and glaucoma. Anatomical systems including the optic disc, blood vessels, and pathological retinal systems including microaneurysms, hemorrhages, exudate, and abnormality types are segmented in this study. The effect of glaucoma in diabetic human beings is discussed, in addition to the evaluation of glaucoma using standards including cup to disc ratio, blood vessels, and anomaly type.

Considerations of anatomical and pathological structures are categorized in the assessment of diabetic retinopathy to identify and categorize DR stages in fundus images. The prime anatomy constitution optic disc is discovered using a modified region growing method that considers the orientation of the image pixel to avoid the drawbacks of prior methods.

The blood vessel is depicted in the anatomy structure that follows. In DR diagnosis, precise segmentation of blood arteries is critical as it might contribute to the creation of aberrant lesions under high pressure. Mathematical morphological operations with distinct structuring elements are used to detect diseased structures such as microaneurysm, hemorrhage, and exudate. Microaneurysms are a common early indication of DR, and they can be discovered by morphologically sealing the green channel fundus image, binarizing it, and using the extended minima transformation. For further processing into normal and abnormal classification, anatomical and pathological structures are segmented. Glaucoma, like diabetic retinopathy, is a retinal disorder in which the OD and OC shapes alter as the fluid pressure level rises. As a result, glaucoma analysis based on OD and OC segmentation is critical for early identification. The OD and OC were segmented in this study utilizing two separate channels, red for the OD and green for the OC, and mathematical morphological techniques were used to segment them. For abnormality categorization of fundus images, the CDR and then blood vessels at inferior superior nasal temporal were used as characteristics given to the NN classifier.

References

[1] A.K. Khurana, Comprehensive Ophthalmology, New Age International (P) Limited Publishers, 2007.

[2] Guyton, Hall, Text Book of Medical Physiology, tenth ed., Elsevier Saunders, 2006.

[3] D. Kourkoutas, I.S. Karanasiou, G.J. Tsekouras, M. Moshos, E. Iliakis, G. Georgopoulos, Glaucoma risk assessment using a non-linear multivariable regression method, Computer Methods and Programs in Biomedicine 108 (3) (2012) 1149–1159.

[4] J.C.H. Lin, Y. Zhao, P.J. Chen, M. Humayun, Y.C. Tai, Feeling the pressure A parylene based intra ocular pressure sensor, IEEE Nanotechnology Magazine 6 (3) (2012) 8–16.

[5] J.L. Olson, M. Asadi-Zeydabadi, R. Tagg, Theoretical estimation of retinal oxygenation in chronic diabetic retinopathy, Computers in Biology and Medicine 58 (2015) 154–162.

[6] Y. Xu, D. Xu, S. Lin, J. Liu, J. Cheng, C.Y. Cheung, T. Aung, T.Y. Wong, Sliding window and regression based cup detection in digital fundus images for glaucoma diagnosis, Medical Image Computing and Computer-Assisted Intervention 14 (3) (2011) 1–8.

[7] G.D. Joshi, J. Sivaswamy, S.R. Krishnadas, Optic disk and cup segmentation from monocular color retinal images for glaucoma assessment, IEEE Transactions on Medical Imaging 30 (6) (2011) 1192–1205.

[8] S.A. Ramakanth, R.V. Babu, Feature match: a general ANNF estimation technique and its applications, IEEE Transactions on Image Processing 23 (5) (2014) 2193–2205.

[9] J. Cheng, J. Liu, Y. Xu, F. Yin, D.W. Wong, N.M. Tan, D. Tao, C.Y. Cheng, T. Aung, T.Y. Wong, Superpixel classification based optic disc and optic cup segmentation for glaucoma screening, IEEE Transactions on Medical Imaging 32 (6) (2013) 1019–1032.

[10] J. Nayak, U.R. Acharya, P.S. Bhat, N. Shetty, T.C. Lim, Automated diagnosis of glaucoma using digital fundus images, Journal of Medical Systems 33 (5) (2009) 337–346.

[11] C.Y. Ho, T.W. Pai, H.T. Chang, H.Y. Chen, An atomatic fundus image analysis system for clinical diagnosis of glaucoma, in: Proceedings of IEEE International Conference on Complex, Intelligent and Software Intensive Systems (CISIS), 2011, pp. 559–564.

[12] S. Dua, U.R. Acharya, P. Chowriappa, S.V. Sree, Wavelet based energy features for glaucomatous image classification, IEEE Transactions on Information Technology in Biomedicine 16 (1) (2012) 80–87.

[13] J. Cheng, F. Yin, Wong, D.W.K. Member, D. Tao, J. Liu, Sparse dissimilarity-constrained coding for glaucoma screening, IEEE Transactions on Biomedical Engineering 62 (5) (2015) 1395–1403.

[14] L.M. Ibrahim, Anomaly network intrusion detection system based on distributed time-delay neural network (DTDNN), Journal of Engineering Science & Technology 5 (4) (2010) 457–471.

[15] J. Sivaswamy, S.R. Krishnadas, A. Chakravarty, G.D. Joshi, Ujjwal, T.A. Syed, A comprehensive retinal image dataset for the assessment of glaucoma from the optic nerve head analysis, JSM Biomedical Imaging Data Papers 2 (1) (2015) 1004–1010.

[16] V. Vijaya Kumari, N. Suriya Narayanan, Diabetic retinopathy early detection using image processing techniques, International Journal on Computer Science and Engineering 2 (2) (2010) 357–361.

[17] F. Berrichi, B. Zohra, Mohamed, Automated Diagnosis of Retinal Images Using the Support Vector Machine SVM, Faculte Des Science, Department of Informatique, USTO, Algerie, 2009.

[18] N. Singh, R.C. Tripathi, Automated early detection of diabetic retinopathy, using image analysis techniques, International Journal of Computers and Applications 8 (2010) 18–23.

[19] P.J. Sargunar, R. Sukanesh, Exudates detection and classification in diabetic retinopathy images by texture segmentation methods, International Journal of ecent Trends in Engineering 2 (4) (2009) 148–150.

[20] J. Goh, L. Tang, G. Saleh, L., Al Turk, Y. Fu, A. Browne, Filtering Normal Retinal Images for Diabetic Retinopathy Screening Using Multiple Classifiers, International Conference on Information Technology and Applications in Biomedicine, 2009, pp. 1–4.

[21] N. Patton, T.,M. Aslam, T. MacGillivray, I.,J. Deary, B. Dillon, R.,H. Eikelboom, Y. Kanagasingam, I. ,J. Constable, Retinal image analysis: concepts, applications and potential, Progress in Retinal and Eye Research 25 (2006) 99–127.

[22] Y. Huajun, Z. Ming, L. Jyh-Charn, Fractal-based automatic localization and segmentation of optic disc in retinal images, International Conference of the Engineering in Medicine and Biology Society (2007) 4139–4141.

[23] C. Muramatsua, N. Toshiaki, A. Sawadac, Y. Hatanakad, T. Haraa, T. Yamamotoc, H. Fujita, Automated segmentation of optic disc region on retinalfundus photographs: comparison of contour modeling and pixel classification methods, Computer Methods and Programs in Biomedicine 101 (2011) 23–32.

[24] K. Shilpa, K. Krishna, Image segmentation and region growing algorithm, International Journal of Computer Technology and Electronics Engineering 2 (1) (2012) 103–107.

[25] A.A. Ajeesha, A. Kumar, Efficient optic disc segmentation and peripappilary atropy detection in digital fundus images, International Journal of Innovative Research in Science and Technology 3 (4) (2016) 213–222.

[26] M.A. Palomera-Perez, M.E. Martinez-Perez, H. Benitez-Perez, J.L. Ortega- Arjona, Parallel multiscale feature extraction and region growing: application in retinal blood vessel detection, IEEE Transactions on Information Technology in Biomedicine 14 (2010) 500−506.

[27] W. Amna, W. Zahra, A. Usman, M. Arslan, Shaukat, Removal of false blood vessels using shape based features and image impainting, Journal of Sensors 2015 (2015) 1−13.

[28] B. Srilatha, V. Malleswara rao, Extraction of blood vessels and exudates from retinal image using image processing algorithms, International Journal of Scientific Engineering and Research 6 (5) (2015) 1652−1656.

[29] M. Kaur, R. Talwar, Automatic extraction of blood vessel and eye retinopathy detection, European Journal of Advances in Engineering and Technology 2 (2015) 57−61.

[30] T. Teng, M. Lefley, D. Claremont, Progress towards automated diabetic ocular screening: a review of image analysis and intelligent systems for diabetic retinopathy, Medical & Biological Engineering & Computing 40 (2002) 2−13.

[31] P. Kahai, K.R. Namuduri, H. Thompson, A decision support framework for automated screening of diabetic retinopathy, International Journal of Biomedical Imaging 2006 (2006) 1−8.

[32] A.A. Purwita, K. Adityowibowo, A. Dameitry, M.W.S. Atman, Automated microaneurysm detection using mathematical morphology, International Conference on Instrumentation, Communication, Information Technology and Biomedical Engineering, 2011.

[33] C. Aravind, M. Poonibala, S. Vijayachitra, Automatic detection of microaneurysms and classification of diabetic retinopathy images using SVM technique, International Journal of Computer Application (2013) 18−22. ICIIIOSP-2013.

[34] G.B. Kande, T.S. Savithri, P.V. Subbaiah, M.N. Tagore, Detection of red lesions in digital fundus images, in: IEEE International Symposium on Biomedical Imaging: From Nano to Macro, 2009, pp. 558−561.

[35] G.B. Kande, S.S. Tirumala, P.V. Subbaiah, Automatic detection of microaneurysms and hemorrhages in digital fundus images, Journal of Digital Imaging (2010) 430−437.

[36] R. Inbarathi, R. Karthikeyan, Classification of splat and GLCM features in fundus images for retinal hemorrhage detection, International Journal of Advanced Research in Computer Science and Software Engineering 4 (2) (2014) 656−662.

[37] S. Siva, D. Raja, S. Vasuki, D. Rajesh Kumar, Performance analysis of retinal image blood vessel segmentation, Advanced Computing: An International Journal 5 (2014) 17−23.

Further reading

[1] W.L. Yun, U.R. Acharya, Y.V. Venkatesh, C. Chee, L.C. Min, E.Y.K. Ng, Identification of different stages of diabetic retinopathy using retinal optical images, Information Sciences 178 (1) (2008) 106−121.

[2] B. van Staal, J.J. Abramoff, M.D. Niemeijer, M. Viergever, M.A. Ginneken, DRIVE: digital retinal images for vessel extraction, IEEE Transactions on Medical Imaging (2004) 501−509.

[3] A.D. Hoover, V. Kouznetsova, M. Goldbaum, Locating blood vessels in retinal images by piecewise threshold probing of a matched filter response, IEEE Transactions on Medical Imaging 19 (3) (2000) 203−210.

[4] Messidor, Methods to Evaluate Segmentation and Indexing Techniques in the Field of Retinal Ophthalmology, Adcis, 2004, http://www.adcis.net/en/Download-Third-Party/Messidor.html, Accessed: January 13, 2022].

[5] T. Kauppi et al., DIARETDB1: standard diabetic retinopathy database, 2007, http://www.it.lut.fi/project/imageret/diaretdb1/, Accessed: January 10, 2022.

Index

'*Note:* Page numbers followed by "f" indicate figures, "t" indicate tables and "b" indicate boxes.'

Printed in the United States
by Baker & Taylor Publisher Services